C. elegans

The Practical Approach Series

SERIES EDITOR

B. D. HAMES
Department of Biochemistry and Molecular Biology
University of Leeds, Leeds LS2 9JT, UK

See also the Practical Approach web site at **http://www.oup.co.uk/PAS**

★ **indicates new and forthcoming titles**

Affinity Chromatography
Affinity Separations
Anaerobic Microbiology
Animal Cell Culture
(2nd edition)
Animal Virus Pathogenesis
Antibodies I and II
Antibody Engineering
Antisense Technology
★ Apoptosis
Applied Microbial Physiology
Basic Cell Culture
Behavioural Neuroscience
Bioenergetics
Biological Data Analysis
Biomechanics – Materials
Biomechanics – Structures and
Systems
Biosensors
★ C. Elegans
Carbohydrate Analysis
(2nd edition)
Cell-Cell Interactions
The Cell Cycle

Cell Growth and Apoptosis
★ Cell Growth, Differentiation
and Senescence
★ Cell Separation
Cellular Calcium
Cellular Interactions in
Development
Cellular Neurobiology
Chromatin
★ Chromosome Structural
Analysis
Clinical Immunology
Complement
★ Crystallization of Nucleic
Acids and Proteins
(2nd edition)
Cytokines (2nd edition)
The Cytoskeleton
Diagnostic Molecular
Pathology I and II
DNA and Protein Sequence
Analysis
DNA Cloning 1: Core
Techniques (2nd edition)
DNA Cloning 2: Expression
Systems (2nd edition)

C. elegans
A Practical Approach

Edited by

IAN A. HOPE
School of Biology
University of Leeds
Leeds LS2 9JT

OXFORD
UNIVERSITY PRESS

OXFORD

UNIVERSITY PRESS

Great Clarendon Street, Oxford OX2 6DP

Oxford University Press is a department of the University of Oxford
and furthers the University's aim of excellence in research, scholarship,
and education by publishing worldwide in

Oxford New York

Athens Auckland Bangkok Bogotá Buenos Aires Calcutta
Cape Town Chennai Dar es Salaam Delhi Florence Hong Kong Istanbul
Karachi Kuala Lumpur Madrid Melbourne Mexico City Mumbai
Nairobi Paris São Paulo Singapore Taipei Tokyo Toronto Warsaw

and associated companies in Berlin Ibadan

Oxford is a registered trade mark of Oxford University Press

Published in the United States
by Oxford University Press Inc., New York

© Oxford University Press, 1999

Users of books in the Practical Approach Series are advised that prudent
laboratory safety procedures should be followed at all times. Oxford
University Press makes no representation, express or implied, in respect of
the accuracy of the material set forth in books in this series and cannot
accept any legal responsibility or liability for any errors or omissions
that may be made.

A catalogue record for this book is available from the British Library

Library of Congress Cataloging in Publication Data
C. elegans : a practical approach / edited by Ian A. Hope.
(The practical approach series ; PAS/213)
Includes bibliographical references and index.
1. Caenorhabditis elegans.   2. Caenorhabditis elegans—Genetics.
I. Hope, Ian A.   II. Series: Practical approach series ; 213.
QL391.N4C22   1999   592'.57—dc21   99–37600

ISBN 0-19-963739-3 (Hbk)
0-19-963738-5 (Pbk)

Typeset by Footnote Graphics,
Warminster, Wilts
Printed in Great Britain by Information Press, Ltd,
Eynsham, Oxon.

Preface

The considerable level of molecular genetic conservation across the animal kingdom means that most genes from other animal species, including humans, have direct homologues in the genome of *Caenorhabditis elegans*. Laboratories concentrating on these other species frequently learn of a *C. elegans* homologue of their gene of interest, when a DNA sequence they have generated is compared to the sequence databases. Identification of such a *C. elegans* homologue is highly likely, given the recent completion of the *C. elegans* genome project—the first for a multicellular organism. The experimental tractability of the *C. elegans* system then makes probing the function of that gene of interest a very attractive proposition.

But what is *C. elegans*? Many have heard of the species, but few have met the organism. The first chapter of this book provides an introduction to this domesticated nematode worm, and explains why it has become such an important system for the study of such a wide variety of biological topics. Having identified a *C. elegans* gene with homology, it is important to gauge the significance of that homology. So, Chapter 2 describes the *C. elegans* genome project that has generated all this sequence data, before examining the quality and quantity of the gene homologies that have been detected so far. If you are new to the *C. elegans* field, and are thinking of starting work on a *C. elegans* gene, it is crucial to find out all that is already known about that gene first. For example, have mutations been identified previously for this gene? If so, are strains with such mutations available, and where from? Chapter 3 describes the computer databases and Web pages through which such information can be readily obtained.

After gathering all the relevant background information, an informed decision to start working with *C. elegans* may be taken. Most experiments will require *C. elegans* to be maintained within the laboratory, as a subject for study. Therefore, procedures for culturing the worms are dealt with next, in Chapter 4.

The first aim may often be to probe gene function through the study of mutations in the gene of interest. Such mutations may already exist, be characterized genetically, and described in the *C. elegans* genetic map—but not yet identified at the molecular level. Complementation of candidate mutations from the appropriate chromosomal region by transformation with a genomic DNA clone would provide a ready-characterized mutant phenotype, and techniques for *C. elegans* transformation are described in Chapter 5. In the absence of a previously characterized mutation, Chapter 6 deals with the current procedures for the direct inactivation of *C. elegans* genes, i.e. reverse genetics. Whether or not mutations in the gene cause an obvious phenotype, more careful observation may be needed. Various techniques for examining *C. elegans* at high magnification are described in Chapter 7. The nervous system is the most complex tissue in the majority of animals, including *C. elegans*, and genes involved in nervous system development and function are the subject of much research. Therefore, different assays of nervous system function, including electrophysiological, pharmacological, and behavioural assays, with which to evaluate more subtle mutant phenotypes, are dealt with in Chapter 8.

A frequently faster route to obtaining preliminary information about a gene's possible function is to examine the developmental expression pattern, and protocols for this are discussed in Chapter 9. Other routine molecular and cell biological techniques are dealt with in Chapters 10 and 11. And, finally, conventional genetic techniques are addressed in Chapter 12. For example, starting from a mutation in one gene, suppressors or enhancers of a mutant phenotype could be isolated to identify other genes involved in the biological phenomenon under study. It seems appropriate to finish with this chapter, because suitability for genetic analysis is the principal advantage of this species for biological investigation and has the potential to lead the researcher on to important new discoveries using this model system.

A book of this sort should be somewhat open-ended. As an introduction to the power of the system, it would have failed if some of those who found this book useful did not wish to pursue their study of *C. elegans* further.

Leeds   I.A.H.
July 1999

Contents

3. *C. elegans* and the Web

4. Maintenance of *C. elegans*

9. Gene expression patterns 181

Andrew Mounsey, Laurent Molin, and Ian A. Hope

Contributors

ROBERT J. BARSTEAD
Oklahoma Medical Research Foundation, Program in Molecular and Cell Biology, 825 NE 13th Street, Oklahoma City, OK 73104, USA

MARK BLAXTER
Institute of Cell, Animal and Population Biology, Ashworth Laboratories, King's Buildings, University of Edinburgh, West Mains Road, Edinburgh EH9 3JT, Scotland

JONATHAN HODGKIN
MRC Laboratory of Molecular Biology, Cambridge CB2 2QH

IAN A. HOPE
School of Biology, University of Leeds, Leeds LS2 9JT

YISHI JIN
Department of Biology, 325 Sinsheimer, University of California, Santa Cruz, CA 95064, USA

IAIN L. JOHNSTONE
Wellcome Unit of Molecular Parasitology, University of Glasgow, Anderson College, 56 Dumbarton Road, Glasgow G11 6NU, Scotland

SHAWN LOCKERY
Institute of Neuroscience, 1254 University of Oregon, Eugene, OR 97403, USA

JIM D. MCGHEE
Department of Biochemistry and Molecular Biology, Faculty of Medicine, University of Calgary, 3330 Hospital Drive NW, Calgary, Alberta T2N 4N1, Canada

PAUL E. MAINS
Department of Biochemistry and Molecular Biology, Faculty of Medicine, University of Calgary, 3330 Hospital Drive NW, Calgary, Alberta T2N 4N1, Canada

LAURENT MOLIN
School of Biology, University of Leeds, Leeds LS2 9JT

ANDREW MOUNSEY
School of Biology, University of Leeds, Leeds LS2 9JT

MICHEL POTDEVIN
CRBM. CNRS, 1919 Route de Mende. 34293 Montpellier, France

RALF SCHNABEL
Institut fuer Genetik, Technische Universitaet Braunschweig, Spielmannstr. 7, D-38106 Braunschweig, Germany

LINCOLN STEIN
CRBM. CNRS, 1919 Route de Mende. 34293 Montpellier, France

THERESA STIERNAGLE
University of Minnesota, 250 Biological Sciences Centre, 1445 Gortner Avenue, St Paul, MN 55108–1095, USA

DANIELLE THIERRY-MIEG
CRBM. CNRS, 1919 Route de Mende. 34293 Montpellier, France

JEAN THIERRY-MIEG
CRBM. CNRS, 1919 Route de Mende. 34293 Montpellier, France

JAMES H. THOMAS
Department of Genetics, Box 357360, University of Washington, Seattle, WA 98195–7360, USA

Abbreviations

5-HT	5-hydroxytryptamine, serotonin
AC	anchor cell
ACh	acetylcholine
ASEL	left member of the ASE class of neurons
ASER	right member of the ASE class of neurons
ATA	aurintricarboxylic acid
BSA	bovine serum albumin
CCD	closed circuit device
cGMP	cyclic 3′, 5′ guanosine monophosphate
CHEF	contour-clamped homogeneous field electrophoresis
cM	centiMorgan
DA	dopamine
DABCO	diazabicyclo[2.2.2]octane
daf	dauer formation (gene)
DAPI	diamidinophenolindole
ddH$_2$O	double-distilled water
DEB	diepoxybutane
Df	deficiency, chromosomal rearrangement
DIC	differential interference contrast
DMF	dimethylformamide
DMP	defecation motor programme
DMSO	dimethyl sulfoxide
DNase	deoxyribonuclease
Dp	duplication, chromosomal rearrangement
d.p.i.	dots per inch
ds	double-stranded
DTC	distal tip cell
DTT	dithiothreitol
E-64	*trans*-epoxysuccinyl-L-leucylamido-(4-guanidino)butane
EDTA	ethylenediaminetetraacetic acid
EGF	epidermal growth factor
EGTA	ethylene glycol-bis-(β-aminoethyl ether)-tetraacetic acid
EM	electron microscopy
EMC	expulsion muscle contraction (aka Exp)
EMS	ethyl methanesulfonate
EPG	electropharyngeogram
EST	expressed sequence tag
Exp	expulsion muscle contraction (aka EMC)
F$_1$	first filial generation

F_2	second filial generation
F_3	third filial generation
FGF	fibroblast growth factor
FITC	fluorescein isothiocyanate
FUdR	fluorodeoxyuridine
GABA	γ-aminobutyric acid
GCG	Genetics Center Group
GFP	green-fluorescent protein
glr	glutamate-receptor (gene)
GLU	glutamate
him	high incidence of males (gene)
HMM	hidden Markov model
HSN	hermaphrodite specific neuron
I_{ca}	inward Ca^{2+} current
I_k	inactivating, outward potassium current
IPSP	inhibitory post-synaptic potential
kb	kilobase
L1–4	larval stages 1–4
LG	linkage group
Mb	megabase
MCS	multiple cloning sites
Muv	multi-vulva
NBT	4-nitro blue tetrazolium chloride
NGM	nematode growth medium
NIH	National Institutes of Health
NIH NCRR	National Institutes of Health National Center for Research Resources
NIMH	National Institute of Mental Health
NLS	nuclear localization signal
NMDA	*N*-methyl-D-aspartic acid
NMJ	neuromuscular junction
p.f.u.	plaque-forming units
p.s.i.	pounds per square inch
P_0	parental generation
PAGE	polyacrylamide gel electrophoresis
PBS	phosphate-buffered saline
PCR	polymerase chain reaction
PEG	polyethylene glycol
PICR	PCR identification of chemically induced rearrangements
PITR	PCR identification of transposable-element induced rearrangements
PMSF	phenylmethylsulfonyl fluoride
Poly-dIdC	poly deoxyinosine deoxycytidine
R	roentgen

RACE	rapid amplification of cDNA ends
RFLP	restriction fragment length polymorphism
RNAi	RNA-mediated interference
RNase	ribonuclease
RSB	Research Services Branch
rtPCR	reverse transcriptase PCR
SDS	sodium dodecyl sulfate
SIT	silicon intensified target
SL	spliced leader
SSC	standard saline citrate
SSRI	serotonin-specific re-uptake inhibitor
STS	sequence-tagged site
t.s.	temperature-sensitive
Tc1	*C. elegans* transposable element 1
TGF-β	transforming growth factor-beta
TMP	4,5',8 trimethylpsoralen, trioxsalen
URL	universal resource locator
WBG	*Worm breeders'gazette*
X-gal	5-bromo-4-chloro-3-indolyl-β-D-galactoside
X-phosphate	5-bromo-4-chloro-3-indolyl-phosphate
YAC	yeast artificial chromosome

1

Background on *Caenorhabditis elegans*

IAN A. HOPE

1. Introduction

The number of researchers using *Caenorhabditis elegans* as a model system for biological research has grown steadily over the past three decades (see *Figure 1*). But interest in this species has suddenly soared in response to the success of the *C. elegans* genome project (1). The genome project was effectively completed in 1998, making *C. elegans* the first multicellular organism for which the genome has been sequenced. The genome sequence data have emphasized the considerable conservation of biological mechanisms across the animal kingdom that has become apparent through the molecular genetic characterization of biological processes. This conservation confirms the value of *C. elegans* as a model system, which has contributed and will continue to contribute significantly to our biological understanding of all animals including humans. *C. elegans* was deliberately selected as a model system (2), primarily because it possessed particularly favourable characteristics for laboratory study and also because it was assumed that similar mechanisms would direct equivalent biological processes in all animal species. This selection has now been shown to be fully justified.

The genome sequence data is providing direct links between laboratories studying many aspects of vertebrate research and those investigating *C. elegans*. As a vertebrate gene is cloned, interrogation of genome sequence databases reveals, more often than not, that there is a direct *C. elegans* homologue. Indeed, because of the advanced state of the *C. elegans* genome project compared to those for other animal species, *C. elegans* is frequently the only species for which a gene homologue may have been previously identified. The possibility of exploring the biological function of that gene and the gene product in a model system, for which progress can be rapid, is attractive. But before taking such an approach, a general appreciation of what is feasible in that system and what is involved for a particular approach is essential. This book should provide all the essential background information for the use of *Caenorhabditis elegans* as a model system.

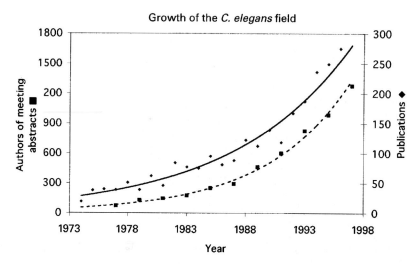

Figure 1. This figure shows the growth in research on *C. elegans* since 1974, the year of the publication (1) considered to herald the use of this species as a model system. The numbers of publications are those added to the CGC (*C. elegans* Genetics Centre) bibliography each year. (Figures kindly provided by Theresa Stiernagle.) The numbers of authors of talk and poster abstracts for presentations at the biennial International *C. elegans* meetings (Cold Spring Harbor, NY up to 1989, Madison, WI since 1991) were derived using the Query builder of the *C. elegans* database ACeDB (see Chapter 3).

2. What is *C. elegans*?

2.1 Natural history

Caenorhabditis elegans is a free-living nematode worm that lives in the soil, across most of the temperate regions of the world, feeding on micro-organisms. With the only requirements for growth and reproduction being a humid environment, ambient temperature, atmospheric oxygen, and bacteria as food, *C. elegans* is particularly cheap and easy to maintain in the laboratory (see *Figure 2*) (see Chapter 4). The adults are only 1 mm long and therefore, although little space is needed to maintain the worms, a microscope is generally required for routine handling. *C. elegans* worms will rarely have been noticed outside the laboratory, despite their prevalence, because of the lack of a direct impact on larger organisms (which announces the presence of parasitic nematodes) and their small size.

For some, 'worm' conveys the notion of a repulsive organism. But *C. elegans* worms are very small and, in the laboratory, confined by water tension to the surface of an agar plate. Movement is restricted to gentle sinusoidal waves almost entirely within a single, dorsoventral plane. They exhibit no obvious smell and their glass-like appearance at high magnification is quite attractive.

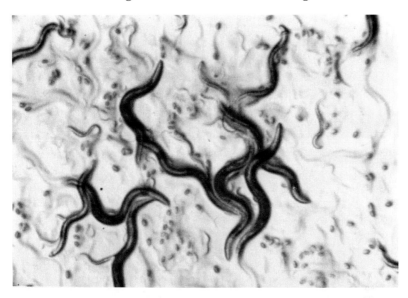

Figure 2. A *C. elegans* culture feeding on bacteria growing on the surface of an NGM agar plate (see Chapter 4). Adults, larvae, and eggs are apparent.

Undergraduates have actually been known to express concern for some of the *C. elegans* mutants they have been given to study!

2.2 Anatomy

There are two sexes, male and hermaphrodite. The hermaphrodite produces both sperm and oocytes and can reproduce by self-fertilization, without mating. The male produces only sperm, and so to reproduce must mate with a hermaphrodite. After mating, male sperm outcompete the hermaphrodites' own sperm in fertilizing the oocytes and most progeny are then the result of cross-fertilization. Cross-fertilization produces males and hermaphrodites in equal proportions, whereas self-fertilization produces only hermaphrodites. This sexual organization has particular value for genetic analysis (see Chapter 12) and was one of the main reasons that *C. elegans* was selected for study.

Both sexes have the same general anatomy (3, 4) and are of similar size, with the adult male being slightly shorter and thinner than the hermaphrodite (see *Figure 3*). The mouth is at the tip of the head, while the anus of the herm-aphrodite and the cloaca of the male are ventral, near to the posterior of the worm. The gut, consisting of a pharynx and an intestine, runs straight from head to tail and is obvious under a microscope because of the transparency of the animal. The pharynx is composed of 20 muscle cells, 20 nerve cells, and 18 epithelial cells arranged with tri-radiate symmetry and enclosed in a basement membrane (5). Through co-ordinated muscular contraction, the pharynx is

3

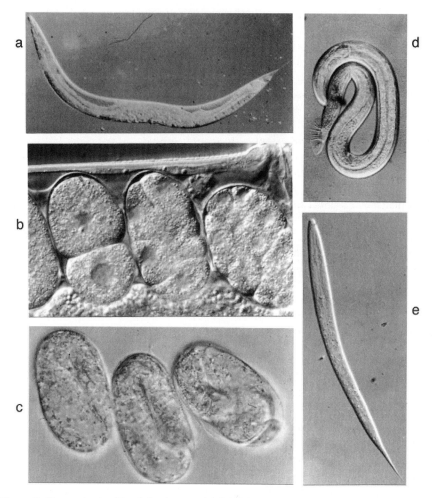

Figure 3. Stages in the life of *C. elegans*. (a) Adult hermaphrodite (100 × magnification). The head is to the left and the tail is to the right. The lumen of the intestine can be seen running most of the length of the worm. Early embryos are located towards the centre of the worm with the midventral vulva facing down. Oocytes, with a square outline and prominent nucleus, are lined up both anteriorly and posteriorly. (b) Early embryos developing in the uterus of a hermaphrodite (1000 × magnification). The two-cell embryo, to the left of centre, has a smaller cell P_1 with a more prominent nucleus, while the nuclear envelope has already broken down in the larger anterior cell, AB, the first of these two cells to divide. The mother's vulva is in the top right of the panel. (c) More mature embryos, laid several hours earlier, at different stages of elongation (1000 × magnification). The embryo on the right is fully elongated. (d) Adult male (200 × magnification). The outline of the double-bulbed pharynx can be seen in the head. The finger-like, sensory rays, in the fan of the tail, contrast strongly with the simple tail spike of the hermaphrodite and larvae. (e) An L1 larva (400 × magnification).

responsible for the ingestion of bacteria, crushing the bacteria in the grinder in the terminal bulb and passing the food into the intestine through the pharyngeal–intestinal valve. The regular flexing of the grinder can be easily observed. The intestine is a single-cell thick tube of apparently simple structure: 20 cells organized in nine rings with all but the most anterior arranged in pairs.

The body wall is separated from the gut by the pseudocoelom and presents a uniform exterior. The single-cell thick hypodermis makes up the bulk of the body wall and secretes the collagenous cuticle that covers the surface of the worm. Most of the hypodermal cells are multinucleate, with one particularly large hypodermal cell occupying most of the dorsal surface at hatching. Through post-embryonic cell-fusion events, this cell also comes to occupy most of the ventral surface in the adult.

Beneath the hypodermis, mononucleate, body-wall muscle cells are organized in four longitudinal bands, running the length of the worm in a subdorsal or subventral position. Each band is two cells wide, the cells having a rhomboid shape. There are 24 muscle cells in each band, apart from the left-ventral band which has 23. These muscle cells send processes to nerve cells, for neuromuscular synapse formation, rather than the more typical converse arrangement. Co-ordinated contraction of the subventral muscle blocks versus the subdorsal muscle blocks generates the waves in body shape that propel the worm on its side along a substrate. Slightly more elaborate movements are possible in the head region, for foraging behaviour, and in the tail region of the male, for mating.

The hermaphrodite's 302 nerve cells (6) are concentrated in the circum-pharyngeal nerve ring, in the ventral nerve cord, and in the tail, with sensory cells concentrated in the head. The nervous system co-ordinates routine behaviour and response to environmental stimuli, and is discussed in detail in Chapter 8. *C. elegans* can sense temperature, a variety of mechanical stimuli, and a wide range of chemical attractants and repellents. The male has an extra 79 nerve cells, and the differences in the nervous systems of the two sexes can be related to differences in sexual behaviour. The connectivity of the entire nervous system has been determined by its reconstruction from serial-section electron micrographs. The simple nervous system was one of the criteria behind the original selection of *C. elegans* as a model system, with the prediction that this would facilitate our understanding of animal behaviour.

The sexes differ in the arrangement of the gonad and in the tail, these differences arising predominantly during post-embryonic development (7, 8). The gonad of both sexes is a simple tubular structure in which the germline replicates through nuclear division within a syncytium. Maturation of the germline, with formation of individual germ cells enclosed in a membrane, occurs with passage along the tube towards the single exterior opening. *C. elegans* sperm cells are amoeboid, like those of other nematodes, and do not have the flagellated morphology more typically associated with sperm cells. In males, the gonad is a single tube opening at the cloaca, close to the tail. The hermaphrodite gonad has two symmetrical lobes (one running towards the anterior, the other towards the posterior), both reflexed back toward the centre of the animal, with the single opening in the uterus located mid-ventrally at the vulva. Fertilization occurs in the spermatheca between the oviduct and the uterus. The hermaphrodite has a set of uterine and vulval muscle cells responsible for egg-laying. In contrast to the simple spike of the hermaphrodite tail, the male tail is an elaborate structure with complex musculature and associated nerves needed for mating. Those new to *C. elegans* rely on the differences in the tail to distinguish the sexes.

There are a few other minor anatomical structures. The excretory/secretory system (9) is thought to be responsible for osmoregulation, and consists of an excretory duct cell, excretory pore cell, excretory gland cell, and the excretory cell, all connected together. The first three cells and the body of the excretory cell are in the body wall, close to the excretory pore, an opening to the ventral surface towards the posterior of the head region. The excretory cell has lateral processes, the excretory canals, which reach to the tip of the head and the tail. And finally there are coelomocytes (six in the hermaphrodite, five in the male), distributed in the pseudocoelom and of uncertain function.

2.3 Life-cycle

Under standard laboratory conditions, growth of *C. elegans* is rapid. The entire life-cycle, from an egg to an adult producing more eggs, takes just 3.5 days at 20°C (see *Figure 3*). (Ranging from less than 3 days at 25°C to 6 days at 15°C. (10)) Population growth is fastest at 20°C, with brood sizes of more than 300 produced over a 4-day period.

Development from fertilization to hatching is referred to as embryogenesis; this generates the first larval stage, which has a similar general structure to the adult, only smaller, being just 250 μm long. Post-embryonic development involves growth through four larval stages (L1 to L4) before the final moult to produce the adult. At 20°C, embryogenesis takes 14 h, the first few hours within the uterus of the hermaphrodite, with the larval stages punctuated by successive moults at 29, 38, 47, and 59 h post-fertilization. Growth during post-embryonic development is continuous.

In the absence of food and at a high enough population density, an alternative stage, the dauer, is formed at the second moult instead of the normal L3. The dauer is specialized for surviving adverse conditions, such as desiccation. Under a dissecting microscope the dauer appears thinner than an L3 larva, remains still most of the time, but moves faster than an L3 when disturbed. The dauer can remain viable for several months and, when food becomes available, moults to become a normal L4 larva.

2.4 Development

There are two phases to embryogenesis (3). During the first 6 h, mitotic cell divisions generate the cells that will make up the first larval stage: 558 in the hermaphrodite and 560 in the male. Initial cell divisions generate cells of unequal size, the so-called founder cells, AB, E, MS, C, and D, as well as the germline P_4. Within each founder-cell lineage, cell divisions are, at least initially, synchronous and produce cells of the same size. E gives rise to all of and only the intestinal cells. D only gives rise to body-wall muscle cells. The other 'founder cells' are not true founder cells as they produce many different cell types. Gastrulation begins at the 28-cell stage, with the two cells of the E lineage moving to the interior of the embryo from the ventral surface. These are followed by P_4, and then the descendants of MS, D, and C. But, on the whole, cells remain in the same relative positions as when they were born, during this initial phase of embryogenesis.

In the second phase of embryogenesis, from 6 h until hatching, morphogenesis turns the ovoid of cells into the larval form. Circumferential actin filaments squeeze the embryo into the elongated shape. Elongation proceeds through the comma, 1.5-fold, 2-fold, 3-fold, and fully elongated stages, with formation of the L1 cuticle fixing the final form. Cellular differentiation occurs during this period of embryo elongation, and as nerve and muscle cells differentiate the embryo begins to first twitch and then roll within the confines of the egg. The apparently, fully formed L1 continues to writhe for a few hours before hatching.

During post-embryonic development (7, 8) the number of somatic cell nuclei increases to 959 in the hermaphrodite and 1031 in the male, and the germline proliferates to fill the gonad as it is formed. These extra cells are generated from a few key blast cells. At hatching, the gonad primordium consists of four cells: Z2 and Z3 are the germline cells and Z1 and Z4 give rise to all the somatic cells of the mature gonad. Sex-specific musculature arises from the single blast cell, M. While the hermaphrodite vulva is derived from three cells in the ventral hypodermis, the male tail is formed from the descendants of three pairs of lateral hypodermal cells, two ventral hypodermal cells and four rectal cells. Post-embryonic development does not involve a metamorphosis, as all the extra structures are formed while maintaining the same overall structure that was generated during embryogenesis.

The simplicity of the animal should be apparent from the anatomical and developmental description given above. The precision of this account is possible because every individual of the species develops at the cellular level in an almost identical fashion. The variation in normal development is restricted to just 11 pairs of lineally related cells. Communication between these pairs of cells results in one cell of each pair following one fate and the other cell of the pair following an alternative fate, but which cell follows which fate varies between organisms.

This consistency of development, in addition to the complete transparency of the animal at all stages of development and the relatively small number of cells making up the fully formed individual, made the complete description of the developmental cell lineage for *C. elegans* possible (3, 4). The cell lineage was determined by following individual cells in live, unstained specimens using differential interference contrast microscopy (see Chapter 7). This cell-lineage description includes the time and orientation of each mitotic cell division as well as the origin and fate of each cell created. Each cell has a unique name, usually a single capital letter for key blast cells. Blast-cell progeny names are derived from the name of the blast cell from which they originated, with additional lower case letters according to the pattern of cell divisions by which they were generated. So, for example, Ca is the anterior daughter of the founder cell C and Cap is the posterior daughter of Ca. Some cells, such as those in the early embryo, are very easy to identify in the live animal, whereas others require considerable experience.

2.5 Evolutionary relationship

How close a model does the nematode *C. elegans* provide for a particular biological topic of study? In part, the answer to this question depends on the evolutionary relationship of *C. elegans* to the species of interest. The nematodes are a deep branch within the evolution of the animal kingdom. Traditional considerations placed the base of this branch close to the division between the deuterostomes and protostomes, i.e. the separation of the insect and vertebrate lineages. There has been some disagreement as to whether the nematodes separated off before the deuterostome/protostome division or branched off from the protostome lineage after the deuterostome/protostome division. Recent molecular studies favour the latter, although these studies also emphasized the difficulty in resolving the relationship between nematodes, insects, and vertebrates (11).

Nevertheless, *C. elegans* is clearly an animal with cell types, such as muscle cells and nerve cells, of similar organization to the equivalent cells of vertebrates. By the time nematodes had branched away from the rest of the animal kingdom, most of the basic mechanisms necessary for the development and functioning of a complex animal may have already evolved. Further evolution of more complex form may have involved mainly the duplication and modification of pre-existing mechanisms. Many biological mechanisms are so well conserved across the animal kingdom that genes can functionally substitute in gene-exchange experiments between *C. elegans* and vertebrates (see, for example, ref. 12). Of 84 human disease genes that have been cloned, 25 have direct orthologues in the *C. elegans* genome and a further 43 have considerable similarity with nematode genes (11). The degree of similarity is perhaps even greater than many of us, who chose to work with *C. elegans* believing that it would be a good model system for improving our understanding of human biology, had expected.

C. elegans is also a model for studies of parasitic nematodes, of major significance in agriculture and medicine. Morphological similarities of the animals in this phylum have meant that the evolutionary relationships within the nematodes have been very difficult to unravel. Furthermore, the Nematoda is an ancient phylum, within which long evolutionary separation is not obviously apparent from the degree of morphological variation. Recent molecular evidence may provide a sounder basis for evaluating how close a model *C. elegans* is for our studies of particular plant and animal parasitic nematodes (13).

3. How can *C. elegans* be studied?

The prospect of moving into a completely new field may be off-putting for those contemplating using *C. elegans* as a model system to further study their biological subject of interest. For example, experts in a particular area of vertebrate biology may be unfamiliar with what is already known in the equivalent area for *C. elegans*. Furthermore, although there are many advantages to using *C. elegans* for laboratory study, like other model systems there are certain disadvantages and therefore some approaches are technically easier or faster than others. Standard, computer-based literature searches can provide important background information, but they may not provide links to other important items of knowledge that would only be apparent to those already working in the field. Fortunately, however, access to the *C. elegans* field is facilitated by the existence of ACeDB, a key computer database, central to the field, into which information on this organism is channelled and through which the body of knowledge concerning *C. elegans* can be explored (see Chapter 3).

ACeDB (14) grew out of the CONTIG9 (15) database originally set up to aid construction of the physical map of the *C. elegans* genome. CONTIG9 included information on the genomic DNA clones covering the genome and the data upon which assembly of this physical map was based. Within ACeDB, genetic data and nucleic acid sequence data have been added to the physical map data, along with all the known connections between these three different views of the *C. elegans* genome. In addition, ACeDB contains information about: cell lineage and development; expressed sequence tags (ESTs); gene structure predictions and homologies; genetic rearrangements; gene expression patterns; metabolic pathways and metabolites; structural proteins and enzymes; laboratories using *C. elegans*; *C. elegans* meetings' abstracts and the *Worm breeders' gazette* articles that are officially unpublished, personal communications between those working in the field. All these different types of data are again interlinked with appropriate connections built into the database.

For example, from an entry point into the field, such as a gene with sequence homology, ACeDB may reveal that a mutation in that gene already exists. ACeDB can then provide a description of a mutant phenotype, from which aspects of gene function may be inferred, and details of the laboratory from which a strain carrying that mutation may be obtained for further study. Many genes are members of gene families within a single genome, possibly with overlapping functions. Because the genome sequence has now been completed for *C. elegans*, all the *C. elegans* genes closely related to the gene that provided the point of entry into the *C. elegans* field will be known, and, hence, ACeDB can provide the information available on the other members of that gene family.

Alternatively, ACeDB may reveal that, although no mutations are known to lie in the gene of interest, there are mutations known which could do, according to their position in the genetic map. Association of a mutation with the gene could be deduced from observing the rescue of a mutant phenotype upon transformation of the mutant strain with the gene in a genomic DNA clone from the *C. elegans* physical map. Both mutant strains and genomic DNA clones are generally freely available. Even though transformation procedures (16) (see Chapter 5) require expensive microinjection equipment and, often, considerable practice to master the technique, transformed strains can be generated by experienced researchers with a few hours of work spread over a couple of weeks. In the absence of a mutation, microinjection of dsRNA (17), in a procedure similar to that for transformation, can be used to specifically disrupt a gene's function.

Transformation with a reporter gene fusion (18, 19) may be used to study the gene's expression pattern (see Chapter 9). The reporter gene fusion would first need to be generated by subcloning promoter regions from the genomic DNA clone into one of the reporter gene vectors specifically designed for *C. elegans*. The anatomical simplicity of *C. elegans* means that expression patterns can be described precisely at a cellular level, although this will usually require some previous experience, particularly for the nervous system. Transparency of the worm at all stages of development means that green-fluorescent protein (GFP) expression patterns are particularly easy to observe in live animals. In addition, 4D-microscope recording systems (20) have been developed, specifically for *C. elegans*, such that embryonic development can be followed more easily within time-lapse images captured at multiple focal planes. *In-situ* hybridization (21) and immunocytochemistry (22) procedures, for examining the distribution of the mRNA or protein produced by the wild-type gene, have been developed and are relatively straightforward. However, antibodies need to be raised first, and the sensitivity and resolution of *in-situ* hybridization may not be as good as that obtained with reporter gene fusions. The reporter gene-fusion approach is probably the best technique to use for initial studies of a gene's expression pattern.

The advantages that *C. elegans* has for genetic approaches may be the major reason for contemplating work with this organism. For example, characterization of a strain carrying a mutation in the gene of interest would be an important objective. If no previously existing mutations are found for a particular gene, then reverse genetic techniques (23, 24) are available for their generation (see Chapter 6). In several laboratories, banks of frozen mutant *C. elegans* strains were set up for screening by the polymerase chain reaction

(PCR) so that strains with transposable element insertions in a target gene could be recovered. Deletions of the gene as a result of imprecise excision of the transposable element may then be screened for, again by PCR. Other banks have been set up using chemically mutagenized worms, which can then be screened for strains with deletions in a gene, directly and in a single step. Each of these screens can be completed in a few weeks, although some genes prove resistant to these approaches. Projects have been initiated with the intention of generating a strain, with a gene deletion, for each gene identified in the genome sequence data. A strain with a particular gene deleted would then be available upon request.

C. elegans is well suited to genetic studies (see Chapter 12). Its small size means that, in a genetic screen, large populations can be examined after mutagenesis for rare mutations. Self-fertilization of the hermaphrodite means that homozygous individuals are generated automatically with no need for sibling-matings, again of great value in genetic screens, and that even strains with severely debilitating mutations can be propagated. Males allow different mutations to be brought together through mating. There is an accepted canonical strain, N2, so that everyone working in the field studies mutations in the same genetic background. The ability to freeze strains in a viable state means that mutant strains can be retained at little cost. Moreover, with the advanced state of the genome project, the molecular cloning of genes defined by mutation is straightforward. However, if a researcher wishes to use *C. elegans* because of a gene homology, genetic studies may only become appropriate after an initial success with other experimental approaches, such as reverse genetics mentioned above.

The study of an individual worm's behaviour is frequently a part of many *C. elegans* projects. Neuromuscular system activity may be monitored in the intact animal from changes in its response to certain stimuli, either mechanical or chemical, or from changes in the pattern of movement across the agar surface. Such activities may be affected by mutations or by the application of specific drugs. Vertebrate homologues are known for many of the proteins needed for the function or assembly of the *C. elegans* neuromuscular system.

Starting with a cloned gene there are several molecular biology techniques that can be applied to *C. elegans*. Probes can be used in *in-situ* hybridization (25) to chromosomal spreads. However, since the chromosomes of *C. elegans* are small, resolution between them depends on using strains with major chromosomal rearrangements. 'Yeast artificial chromosome (YAC) polytene grids' (26), with YAC genomic DNA clones gridded out in order as the inserts are found in the genome, and covering most of the genome, were a by-product of the physical map. Probing such grids is a convenient alternative to chromosomal *in-situ* hybridization. In addition, genotyping can be performed on individual animals by 'single-worm PCR.'

Other experimental approaches, routine with other species, may be difficult or impossible for *C. elegans*. There are no nematode cell lines available. While

C. elegans can be grown in bulk in liquid culture, such preparations will, of course, contain all the different tissues present in the intact animal. The small size of the adult means that dissection of specific tissues is unrealistic, although preparations of the pharynx have been used successfully for electro-physiological studies (27) and careful dissection has allowed patch-clamp recordings to be obtained for individual neurons (28). Electrophysiological work has also been carried out using larger nematode species, which appear to have a similar nervous system organization (29). The small size of the embryo means that grafting experiments used for the study of development in other species are not feasible for *C. elegans*, although some manipulation at the single-cell level is routine. First, individual blastomeres from the early embryo can be isolated and will continue to develop for many hours in culture (30). Second, the relative positions of early blastomeres can be changed by applying pressure to the surface of the egg (31). Third, laser microbeams are used to ablate specific cells in live individuals at embryonic or post-embryonic stages (32).

Most studies with *C. elegans* involve either the whole animal or cloned genes. An appreciation of the feasibility and time-scale of the different experimental approaches is crucial before starting work with *C. elegans*.

4. What biological subjects has *C. elegans* had a key role in?

One of the most prominent impacts of work with *C. elegans* is, of course, with regard to the genome sequencing projects (1). Procedures used in construct-ing the *C. elegans* physical map and in determining the sequence itself have influenced the strategies of genome projects for other species. ACeDB (14), the database developed for handling the *C. elegans* sequence data and in-tegrating it with other information about *C. elegans*, has been adopted by those studying other organisms. In the future, the approaches developed for interpreting the *C. elegans* sequence data may have a major influence on how the data from the sequencing of other genomes is utilized. Moreover, the high level of homology between the worm and humans (11) means that our under-standing of the *C. elegans* data will probably contribute significantly to our comprehension of the human genome sequence. But work with *C. elegans* has already yielded important insights into other areas of biology.

Work with *C. elegans* has had a major impact on our understanding of apoptosis or programmed cell death (33). Genetic observations with *C. elegans* led to our appreciation of the significance of the process for animal develop-ment and in the maintenance of the fully formed animal. The mechanism of programmed cell death appears to be highly conserved across the animal kingdom, and work with the *C. elegans* cell-death genes has been instrumental in our understanding of the process at the molecular level.

Many other aspects of animal development have been investigated in *C. elegans*. Invariant development had suggested that development in this species would be controlled in a cell-autonomous fashion. However, intercellular signalling does have a major role in directing the development of *C. elegans*, and this has been extensively analysed. *C. elegans* genetics has proved to be a powerful approach for dissecting signal transduction pathways by which signals, perceived at the cell surface, are converted into a response within the cell, e.g. in induction of the vulva (34). Members of the important families of transcriptional regulators with fundamental roles in directing animal development, such as the *Hox* genes (35) and the myogenins (36), have been studied in *C. elegans*. UNC-86 was one of the four founding members of the POU class of homeodomain proteins (37), and LIN-11 and MEC-3 were two of the three founding members of the LIM class of homeodomain proteins (38). *C. elegans* proteins involved in the morphogenesis and function of nerve cells and muscle cells were identified through mutations. The cloning of *unc-54*, over 15 years ago, provided the first complete sequence of a myosin heavy chain (39). Now, the genome project is making *C. elegans* the first organism for which entire families of proteins, which have a particular function in the nervous system, such as the chemosensory receptors, will be known. Many other topics in biology, e.g. sex determination (40), ageing (41), etc., have also been studied extensively in *C. elegans*.

Homology revealed in gene sequence is providing the links between *C. elegans* and humans. Biological understanding in *C. elegans* can now be translated more easily into a biological understanding of our own species.

5. Other books about *C. elegans*

This book is primarily intended to be an introductory guide to *C. elegans* for researchers currently working outside the field. There are three other key books that provide more detail about this species. *The nematode,* Caenorhabditis elegans, edited by Wood (42), is now 10 years old, but is a vital background text. C. elegans *II*, edited by Riddle, Blumenthal, Meyer, and Priess (43), was written a couple of years ago as an update of the first book and provides a summary of further developments in the field. And last but not least, Caenorhabditis elegans: *modern biological analysis of an organism*, edited by Epstein and Shakes (44), contains detailed accounts of techniques used in work with *C. elegans*. The *C. elegans* web server (`http://elegans.swmed.edu/`) and a web site for *C. elegans* protocols (`http://www.dartmouth.edu/artsci/bio/ambros/protocols/`) are also valuable sources

References

1. The *C. elegans* sequencing consortium (1998). *Science*, **282**, 2012.
2. Brenner, S. (1974). *Genetics*, **77**, 71.

3. Sulston, J. E., Schierenberg, E., White, J. G., and Thomson, J. N. (1983). *Dev. Biol.*, **100**, 64.
4. Sulston, J. E. and Horvitz, H. R. (1977). *Dev. Biol.*, **56**, 110.
5. Albertson, D. G. and Thomson, J. N. (1976). *Phil. Trans. R. Soc. Lond. B. Biol. Sci.*, **275**, 299.
6. White, J. G., Southgate, E., Thomson, J. N., and Brenner, S. (1986). *Phil. Trans. R. Soc. Lond. B. Biol. Sci.*, **314**, 1.
7. Sulston, J. E., Albertson, D. G., and Thomson, J. N. (1980). *Dev. Biol.*, **78**, 542.
8. Kimble, J. and Hirsh, D. (1979). *Dev. Biol.*, **70**, 396.
9. Nelson, F. K., Albert, P. S., and Riddle, D. L. (1983). *J. UltraStruct. Res.*, **82**, 156.
10. Byerley, R. C., Cassada, R. C., and Russell, R. L. (1976). *Dev. Biol.*, **51**, 23.
11. Mushegian, A. R., Garey, J. R., Martin, J., and Liu, L. X. (1998). *Genome Research*, **8**, 590–98.
12. Vaux, D. L., Weissman, I. L., and Stuart, S. K. (1992). *Science*, **258**, 1955.
13. Blaxter, M. L., De Ley, P., Garey, J. R., Liu, L. X., Scheldman, P., Vierstraete, A., Vanfleteren, J. R., Mackey, L. Y., Dorris, M., Frisse, L. M., Vida, J. T., and Thomas, W. K. (1998). *Nature*, **392**, 71–75.
14. Eeckman, F. H. and Durbin, R. (1995). In Caenorhabditis elegans: *modern biological analysis of an organism* (ed. H. F. Epstein and D. C. Shakes), p. 584. Academic Press, San Diego, CA.
15. Sulston, J., Mallett, F., Staden, R., Durbin, R., Horsnell, T., and Coulson, A. (1988). *CABIOS*, **4**, 125–32.
16. Mello, C. C., Kramer, J. M., Stinchcomb, D., and Ambros, V. (1991). *EMBO J.*, **10**, 3959.
17. Fire, A., Xu, S., Montgomery, M. K., Kostas, S. A., Driver, S. E., and Mello, C. C., (1998). *Nature*, **391**, 806–11.
18. Fire, A., Harrison, S. W., and Dixon, D. (1990). *Gene*, **93**, 189.
19. Chalfie, M., Tu, Y., Euskirchen, G., Ward, W. W., and Prasher, D. C. (1994). *Science*, **263**, 802.
20. Hird, S. N. and White, J. G. (1993). *J. Cell Biol.*, **121**, 1343.
21. Seydoux, G. and Fire, A. (1994). *Development*, **120**, 2823.
22. Albertson, D. (1984). *Dev. Biol.*, **101**, 61.
23. Rushforth, A. M., Saari, B., and Anderson, P. (1993). *Mol. Cell. Biol.*, **13**, 902.
24. Zwaal, R. R., Broeks, A., van Meurs, J., Groenin, J. T. M., and Plasterk, R. H. A. (1993). *Proc. Natl Acad. Sci. USA*, **90**, 7431.
25. Albertson, D. (1984). *EMBO J.*, **3**, 1227.
26. Coulson, A., Kozono, Y., Lutterbach, B., Shownkeen, R., Sulston, J., and Waterston, R. (1991). *Bioessays*, **13**, 413.
27. Raizen, D. M. and Avery, L. (1994). *Neuron*, **12**, 483.
28. Goodman, M. B., Hall, D. H., Avery, L., and Lockery, S. R. (1998). *Neuron*, **20**, 763–72.
29. Stretton, A. O. W., Fishpool, R. M., Southgate, E., Donmoyer, J. E., Walrond, J. P., Moses, J. E. R., and Kass, I. S. (1978). *Proc. Natl Acad. Sci. USA*, **75**, 3493–7.
30. Schierenberg, E. (1987). *Dev. Biol.*, **122**, 452.
31. Priess, J. R. and Thomson, J. N. (1987). *Cell*, **48**, 241.
32. Chalfie, M., Sulston, J. E., White, J. G., Southgate, E., Thomson, J. N., and Brenner, S. (1985). *J. Neuroscience*, **5**, 956–64.

33. Metzstein, M. M., Stanfield, G. M., and Horvitz, H. R. (1998). *Trends Genet.*, **14**, 410.
34. Kornfeld, K. (1997). *Trends Genet.*, **13**, 55.
35. Kenyon, C. J., Austin, J., Costa, M., Cowing, D. W., Harris, J. M., Honigberg, L., Hunter, C. P., Maloof, J. N., MullerImmergluck, M. M., Salser, S. J., Waring, D. A., Wang, B. B., and Wrischnik, L. A. (1997). *Cold Spring Harbor Symp. Quant. Biol.*, **62**, 293–305.
36. Harfe, B. D., Branda, C. S., Krause, M., Stern, M. J., and Fire, A. (1998). *Development*, **125**, 2479–88.
37. Finney, M., Ruvkun, G., and Horvitz, H. R. (1988). *Cell*, **55**, 757.
38. Freyd, G., Kim, S. K., and Horvitz, H. R. (1990). *Nature*, **344**, 876.
39. Karn, J., Brenner, S., and Barnett, L. (1983). *Proc. Natl Acad. Sci. USA*, **80**, 4253.
40. Marin, I. and Baker, B. S. (1998). *Science*, **281**, 1990.
41. Hekimi, S., Lakowski, B., Barnes, T. M., and Ewbank, J. J. (1998). *Trends in Genetics*, **14**, 14–19.
42. Wood, W. (ed.) (1988). *The nematode* Caenorhabditis elegans. Cold Spring Harbor Laboratory Press, NY.
43. Riddle, D. L., Blumenthal, T., Meyer, B. J., and Priess, J. R. (ed.) (1997). C. elegans *II*. Cold Spring Harbor Laboratory Press, NY.
44. Epstein, H. F. and Shakes, D. C. (1995). Caenorhabditis elegans: *modern biological analysis of an organism*. Academic Press, San Diego, CA.

2

The genome project and sequence homology to other species

MARK BLAXTER

1. The *C. elegans* genome project

Caenorhabditis elegans has a small genome (1–3). At 97 megabases (Mb) it is one-thirtieth the size of the human (and other mammalian) genomes. Since *C. elegans* is a metazoan, its genome might be expected to be significantly more complex than those of single-celled eukaryotes such as yeasts and protozoa, yet it is only three times the size of the malaria genome and eight times that of fission yeast. Within this relatively small gene set lie all the instructions for the development and functioning of a fully differentiated animal. These include genes for developmental regulation and embryogenesis, the functioning of complex organ systems, and the integration of the nervous system to control behaviour (4–8).

At the inception of the *C. elegans* genome project, the idea of determining the complete sequence (never mind the structure) of such a vast genome was nearly unthinkable. Thanks to the vision, drive, and sheer hard work of a core group of researchers (in particular John Sulston, Alan Coulson, and Bob Waterston), the dream of having a fully mapped, and fully sequenced genome is complete. The genome project has been central to the success of *C. elegans* as a model organism, as it allows researchers to home in on the molecular basis of observed genetic defects without having to spend years in the forest of conventional, mapless cloning projects (9). The genome sequence yields the ultimate in genotyping and is the basis on which the next generation of 'post-genomics' research will be based. In addition, the finding of a potential homologue of a gene-of-interest in the *C. elegans* genome sequence dataset is often the first inkling a researcher will have that this small nematode might be of interest to them.

The genome project has three main strands. The first is the maintenance and integration of the *C. elegans* genetic map, including molecular genetic markers such as sequence tagged sites and transposon insertion sites. The second is the generation of a fully contiguated physical map of the six chromosomes using a combination of cosmid and yeast artificial chromosome

(YAC) clones (10–12). The third is the sequencing of the whole genome based on a minimum-overlap set of the clones in the physical map (13,14). The three strands are integrated in the analysis and annotation of the sequence in the genome database ACeDB (see Chapter 3), with loci identified by mutation serving to link the genetic and physical maps, and the sequence map serving to inform experiments designed to isolate the mutated loci (15–18).

One of the continuing features of the *C. elegans* field is the open and easy access afforded to all. This is exemplified by the genome project, where sequence, analysis and clone data, and reagents are freely and publicly available (see *Boxes 1* and *2*). This unprecedented openness of access means that the full dataset is there for any researcher to mine and sift for nuggets relevant to their interests and programmes. The open availability of nematode strains and genome project reagents also means that the start-up curve for anyone entering the field is relatively shallow. In this chapter I will describe the physical and sequence maps of *C. elegans*, introduce ways in which this information can be accessed, and give an overview of sequence and functional homologies to other species which can help researchers in transferring their system to the *C. elegans* model. Gene function can then be explored *in vivo* using the powerful techniques available for *C. elegans* as described in the following chapters.

2. The physical map

The *C. elegans* physical map was constructed from cosmid, YAC, and fosmid clones (3, 11). First, approximately 17 000 cosmid clones (insert size ~35 kb) were fingerprinted, and contiguated using fingerprint overlap data (10, 19). Second, selected cosmid clones were mapped to a 3000-clone YAC library (insert size >150 kb to 1 Mb) by hybridization (12). Fosmid clones have also been used to provide coverage of some regions. In the sequencing phase, long-range polymerase chain reactions (PCR) were used to span 44 apparent gaps in clone coverage. The current map has only five 'gaps' (regions not covered by cloned DNAs). Of these gaps, three are between telomeres and the non-repetitive segments of the chromosome, and two are internal.

The cosmid fingerprinting was achieved by restricting each clone with *Hin*dIII, labelling the fragment ends with [^{32}P], and cutting with a second, frequent-cutting enzyme (*Sau*3AI). The labelled fragments were analysed by acrylamide gel electrophoresis, and the resulting autoradiographs digitized. These digital fingerprints were then examined using custom software with subsequent direct inspection of matched fingerprints to identify clones that shared significant overlap, and contigs of overlapping clones were built up. This process resulted in over 600 contigs, some of only a very few clones. These contigs were mapped to the YAC library by selecting marker cosmids spanning each contig and hybridizing them to high-density YAC library filters. This process resulted in a massive reduction in the number of contigs, to the current position where over 99% of the genome is covered by either YAC and/or cosmid clones. The use of two different host–vector systems appears to have been very successful in getting round problems of 'unclonable' DNA: segments of the genome which can be propagated only poorly (if at all) in one system appear to be stable in the other.

Box 1   Access to *C. elegans* genome materials and data[1]

Clones and strains
Physical map clones
   cosmids, YACs, and YAC library high-density gridded filters
      Alan Coulson, Sanger Centre, Hinxton (alan@sanger.ac.uk)
Transgenic strains
   carrying sequenced cosmid clones
      David Baillie, SFU, Vancouver (baillie@sfu.ca)
EST clones
   yj series
      Yuji Kohara, NIG, Japan (kohara@nig.jp)
   wEST series
      AR Kerlavage, The Institute for Genomic Research (arkerlav@tigr.org)
   cm series
      Alan Coulson, Sanger Centre, Hinxton (alan@sanger.ac.uk)

Sequences
Genomic sequence and predicted proteins
   GenBank
      http://www.ncbi.nlm.nih.gov/
   EMBL
      http//www.ebi.ac.uk/
Current genome sequence including unfinished sequence
   Sanger Centre
      http://www.sanger.ac.uk/Projects/C_elegans/Genomic_Sequence.shtml
   Washington University Genome Center
      http://genome.wustl.edu/gsc/gschmpg.html
ESTs
   dbEST
      http://www.ncbi.nlm.nih.gov/dbEST/index.html

[1]The other nematode EST and genome projects have similar arrangements for access to materials (see, for example, the *Brugia* web site at http://helios.bto.ed.ac.uk/mbx/fgn/filgen1.html)

The physical and genetic maps have been anchored to each other by the molecular cloning of genes defining particular loci. There are over 2000 genetic loci defined in *C. elegans*, and for a significant proportion of these (800) the affected gene is known. In addition, many deficiency (large chromosomal deletion) endpoints have been mapped to both genetic and physical maps.

The ambiguities and conflicts in the physical map were resolved as sequencing approached completion. The map is co-ordinated by Dr A. Coulson of the Sanger Centre, Hinxton. The physical map is publicly available in the *C. elegans* genome database ACeDB (see Box 1), and clones from the map are freely available.

3. The expressed genome

The number of visibly mutable loci in *C. elegans* was initially estimated at about 2000 (20). There are currently 2254 'named' loci registered within ACeDB (at July 1999). With the genome sequencing project and extensive additional analysis at the genetic level, it is clear that the actual number of genes is much higher. Yeast has about 6000 genes, and humans are predicted to have between 80 000 and 100 000. *C. elegans* has over 19 000 protein coding genes (3, 4). One way of examining the genes expressed by an organism is to examine them at the mRNA level rather than at the genomic level (21, 22). Analysis of mRNA means that the complications (and extra sequencing work) of introns and intergenic segments and other non-coding DNA can be ignored. This expressed sequence tag (EST) approach has been applied to *C. elegans*, and there are currently three EST datasets available (21–24). Of these, two are relatively small and are derived from cDNA libraries from mixed-stage cultures of nematodes. One was generated at The Institute for Genome Research, Gaithersburg, MA (sequences with 'wEST' designators; 2400 ESTs from 3' and 5' reads) (21), and the other at the Sanger Centre (sequences with a 'cm' designation; 1500 5' read ESTs) (22). Both are single-direction sequencing reads on clones derived from libraries which had been normalized, by subtraction hybridization, and they identify about 1200 different *C. elegans* genes. The third EST dataset is much more extensive (68 000 sequences from 40 000 clones, with both 3' and 5' reads from 75% of clones) and has been generated by Y. Kohara's laboratory (National Institute of Genetics, Mishima, Shizuoka Japan) as part of a global project to identify the expressed genome of the nematode (24) (see *Box 2*). This dataset derives from both adult stage and embryonic cDNA libraries, and has identified about 38% of the genes of *C. elegans* (3). The Kohara EST dataset has been generated by extensive sampling of initially non-normalized libraries. The clones are designated with a 'yk' prefix. As the project has progressed, periodic subtraction hybridization (of abundant sequences) has been performed to reduce the redundancy of the sequencing.

The ESTs have been used by the genome sequencing consortium to confirm gene predictions and predict alternative splicing (about 1% genes are currently predicted to be alternatively spliced) (3). The clones are useful starting points for gene analysis, for example expression in heterologous systems, mutagenesis, etc. All the EST sequences are available in the public sequence databases (see *Box 1*), and also, for the Kohara laboratory's project, as an analysed and annotated dataset on a dedicated World Wide Web site (see *Box 2*). The ESTs are also integrated into ACeDB. EST clones are generally available from the sequencing laboratories.

Box 2   Analysis and documentation of *C. elegans* genome data

Genome Sequencing and genomics
C. elegans genomics and genetics resources
      http://elegans.swmed.edu/genome.shtml
C. elegans at the Sanger centre
      http://www.sanger.ac.uk/Projects/C_elegans/
WUSTL Genome Center
      http://genome.wustl.edu/gsc/gschmpg.html
Sequence-mapped Tc1 transposon insertion sites
      http://www.sanger.ac.uk/Projects/C_elegans/tc1poly/
Transgenic *C. elegans* carrying sequenced cosmids
      http://darwin.mbb.sfu.ca/imbb/dbaillie/cosmid.html
C. elegans repeat families
      http://www.sanger.ac.uk/Projects/C_elegans/repeats/
Introduction to Genefinding in *C. elegans*
      http://www.sanger.ac.uk/Projects/C_elegans/genefinding/
      genefinding_resources.shtml
Introduction to the Kohara lab EST and expression pattern WWW site
      http://www.ddbj.nig.ac.jp/htmls/c-elegans/html/CE_INDEX.html
Keyword search of Kohara EST and expression pattern site
      http://www.ddbj.nig.ac.jp/c-elegans/html/CE_KEYWORD.html

Sequence similarity search engines
NCBI BLAST similarity search servers
      http://www.ncbi.nlm.nih.gov/cgi-bin/BLAST/nph-blast?Jform=0
EMBL similarity search engines
      http://www.ebi.ac.uk/
BLOCKS protein family search engines
      http://blocks.fhcrc.org/blocks_search.html
Pfam protein family search engines
      http://www.sanger.ac.uk/Software/Pfam/
C. elegans BLAST server (Cambridge)
      http://www.sanger.ac.uk/Projects/C_elegans/blast_server.shtml
C. elegans BLAST server (WUSTL)
      http://genome.wustl.edu/gsc/blast/blast_servers.html

Box 2 *Continued*

Automated BLAST searching of *C. elegans* databases
  http://www.sanger.ac.uk/Projects/C_elegans/Autoblast/
BLAST search of the Kohara lab EST dataset
  http://www.ddbj.nig.ac.jp/c-elegans/html/CE_BLAST.html
Parasitic nematode BLAST server
  http://www.ebi.ac.uk/parasites/parasite_blast_server.html
Hidden Markov Models software
  http://hmmer.wustl.edu/

Sequence databases

ACeDB *C. elegans* on the WWW (simple text format)
  http://www.sanger.ac.uk/Projects/C_elegans/webace_front_end.shtml
ACeDB *C. elegans* on the WWW (full graphical format)
  http://webace.sanger.ac.uk/cgi-bin/webace?db=wormace&Continue=Continue&.cgifields=db
Wormpep (predicted proteins from the genome project)
  http://www.sanger.ac.uk/Projects/C_elegans/wormpep/
Retrieving sequences from Wormpep or the *C. elegans* ESTs
  http://www.sanger.ac.uk/Projects/C_elegans/wormpep_fetch.shtml
The *C. elegans* mitochondrion
  http://www.sanger.ac.uk/~dl1/MTDNA_2.0/mt_home.html

Gene expression databases and search engines

betagalactosidase tagged expression patterns (Hope lab expression
  pattern database)
  http://www.personal.leeds.ac.uk/~acedb/Hope/epa.htm
in situ hybridisation expression patterns (Kohara lab expression pattern
  database)
  http://watson.genes.nig.ac.jp:8080/db/index.html

General *C. elegans* sites

Caenorhabditis Genetics Center (strains and bibliography, also the Worm
  Breeders' Gazzette)
  gopher://elegans.cbs.umn.edu:70/1
C. elegans community www site SWMED (curated by Leon Avery)
  http://elegans.swmed.edu/
Worm Labs on the Web
  http://elegans.swmed.edu/Worm_labs/
C. elegans literature search engine (published papers and abstracts from
  the Worm Breeders' Gazzette and meetings)
  http://elegans.swmed.edu/htbin/wbgart/

4. The genome sequence

The sequence of the *C. elegans* genome has been determined by a consortium based at the Sanger Centre, Cambridge, UK, and at the Genome Center, Washington University of St Louis, MI, USA (3, 13, 14, 25). These two sites have each generated about 50 Mb of sequence, analysed, and annotated it. In doing so they have pioneered large-scale, academic sequencing, and *C. elegans* is thus a model for the mechanics of the human genome project. The sequence has been generated using a minimum tiling path of cosmid (2527 clones), YAC (257), and fosmid (113) clones from the physical map as substrate. 44 gap regions are covered by PCR products rather than clones.

The sequencing project was explicitly map- and clone-based, rather than shotgun. A cosmid was selected that had minimum overlap with its neighbours, and the insert was subcloned into M13 phage and plasmid sublibraries of 1–2 kb and 6–9 kb fragments. The longer clones were useful in directed sequencing to close gaps in the first-pass assembled sequence. Randomly selected subclones were then sequenced (~800 per cosmid) and these shotgun sequences used to generate contiguated DNA sequence for the cosmid. The initial shotgun phase was followed by a finishing phase of directed resequencing of clones that span areas of ambiguity or gaps in the sequence. The finished sequence has been determined on both strands, to an average density of 6–8 reads per base. Initially, sequencing concentrated on the cosmid clones, but as these were exhausted, attention turned to gaps in the cosmid coverage. These gaps were covered and sequenced by a combination of strategies including long-range PCR, rescue of joining fragments from YACs, and YAC sequencing.

The project progressed by first sequencing the gene-rich central portions of the autosomes (which were also very well represented in the cosmid libraries), followed by sequencing on the X chromosome (where genes are less dense). The termini of the autosomes and X chromosome were the last segments to be covered, but as they contain only a small proportion of the genes (40% of the DNA but 20% of the genes) the bulk of the protein-coding genes were identified very rapidly. Assessment of the accuracy of sequence determination is made by resequencing clones in the two sites, overlap between neighbouring cosmids, and comparison to sequences deposited from other laboratories studying *C. elegans*: the error rate is estimated at about 1 in 100 000 bases. The project has a rigorous and thorough quality control system, and all the annotated sequence is finished to this same high standard (3).

As sequencing of a clone (cosmid or YAC) approaches completion, the sequence is released directly to the public databases. The genome consortium does not stop there, however, as their goal is to produce a fully annotated genome sequence. The annotation process uses computational tools developed for the project (but of use elsewhere) to examine the genomic sequence and predict encoded features. These tools are at the forefront of genome analysis bioinformatics, and are continually being refined and improved (25). Many are based on algorithms trained to recognize sequence features by using a starting dataset: as the genome sequence reaches completion the predictive power of these algorithms is improved. It is thus likely that the process of revision and refinement of annotation will continue for some time after the 'official' completion of the genome.

5. Sequence analysis and annotation

5.1 Annotation of coding segments

The most important informatics task is the identification of genes. Analysis and annotation is carried out within the genome database ACeDB, with additional external tools used for refinement and confirmation of gene predictions (15, 16, 18). The *C. elegans* genome is relatively gene-rich, with one gene per approximately 6 kb. Genes are closer together in the central gene-rich portions of the autosomes, less dense on the X chromosome, and least dense on the peri-telomeric autosomal arms (4). Introns in *C. elegans* tend to be short (as small as 37 bases) (26), and there are strong, splice-site consensus sequences (particularly the 3′ splice acceptor site) (27). The genome is about 60% AT (1), with introns and intergenic regions being more AT-rich than exons. These features are used to predict genes in the genome sequence, by identifying putative exons and examining their potential to be spliced to form an mRNA with a significant open reading frame. Gene prediction is aided by comparing the genomic sequence to the *C. elegans* EST dataset and to public protein-sequence datasets. If an EST is derived from a genomic segment, it must contain an expressed gene. The ESTs also confirm splicing patterns (including alternative splicing). The Kohara EST dataset includes both 5′ and 3′ (poly(A)) sequence reads from most clones, and these can be used to define the limits of genes. Comparison to public protein-sequence databases highlights regions of the genome which have the potential to encode peptides similar to known sequences. These different modes of information are integrated to produce the final gene prediction. Genes are named after the clone where they are found, and are each given a numerical suffix. Thus predicted genes ZK637.1, ZK637.2, and ZK637.3 all derive from the cosmid ZK637. It is important to note that the genes are not numbered with reference to their arrangement on the cosmid (i.e. ZK637.1 will not necessarily be next to ZK637.2). For genes predicted by GeneFinder, but not reviewed manually, a lower-case suffix is used (e.g. ZK637.a would be a predicted but unverified gene on ZK637).

The genome sequence has also been analysed for the presence of local direct or inverted repeats, the presence of various defined *C. elegans* repeat families (3, 28–30), and other long- and short-range sequence features (see

Box 2). The genome, like other metazoan genomes, is littered with the remains of degenerating transposons and other mobile elements (31, 32) which are identified and tagged. Simple sequence regions are also identified.

For some genes, the task is relatively simple, as both ESTs and protein similarity data are available. For others, where there are no ESTs or informative similarities, prediction is made on the basis of open reading frames, splicing compatibility, and base content bias. Thus for any gene predicted in the genome, it is important for a researcher to assess the strength of the prediction. Are there ESTs? Are there informative similarities? In particular, the gene prediction algorithms are worst at predicting the 5′ and 3′ ends of genes (although they are still very good at even this task). This is to be expected, as, in the absence of EST information, the relative merit of extending the 3′ end of an open reading frame by a few codons by introducing another splicing event is, as yet, impossible to assess objectively. For each gene, therefore, it is quite important to perform some independent assessment of the accuracy of the predictions made, particularly if the gene is otherwise anonymous. The genome database ACeDB retains and displays all the information used in predicting genes (the database similarity searches, EST similarities, potential splice sites, base content) and is an essential workbench on which to examine and confirm predictions. New knowledge of a homologue from a different species may be of value in refining gene structure predictions.

5.2 Other data on the sequence map

The sequence map is the ultimate integration of physical and genetic maps, and is extensively annotated with additional information derived from genome-based projects. The Baillie laboratory in Simon Fraser University, Vancouver, Canada, is generating an extensive genetic toolkit for *C. elegans* (33). Part of this project is the generation of transgenic strains carrying each of the sequenced cosmids. These are freely available, and can be used for phenotype rescue in crosses to strains harbouring mutations at loci that may be covered by the cosmid.

There are several different types of transposon in *C. elegans*, and different strains and isolates have different numbers of transposons in their genomes. Transposon insertion mutagenesis has been, and remains, a powerful tool for generating mutations in target loci, and transposons can be used as molecular genetic markers for mapping and cloning loci of interest. The positions of 439 transposons in the genomes of several strains of *C. elegans* (including the designated wild-type, N2) have been mapped by tag sequencing of transposon–genome junctions and comparing these to the sequence map (312 mapped transposons), and by hybridizing junctional fragments to physical map clones (127 transposons) (34).

6. Searching the *C. elegans* databases for homologues of genes of interest

The first step in many projects initiated on *C. elegans* is now a sequence-based approach to identifying genes of interest. With the complete sequence to hand, it is only necessary to ask the right questions of the databases to isolate putative homologues. The *C. elegans* data are all available in the major public databases (GenBank, EMBL, DDBJ), but a more efficient search can be made using a *C. elegans*-only dataset (see *Box 2*). In order of complexity, the different sorts of searches that can be made are:

(a) simple-sequence similarity searches;

(b) matrix-based similarity searches; and

(c) motif or pattern searches.

One of the major insights emerging from the *C. elegans* project is that the level of genetic homology between species is greater than anticipated, e.g. signal transduction mechanisms appear to have been highly conserved as multigene assemblies (35–37). Thus the epidermal growth factor (EGF), signal-receptor cascade is present in nematodes, as is that for insulin-like growth factors, transforming growth factor-beta (TGF-β), and other systems. Intracellular signalling pathways are also conserved, and have been recruited wholesale to new function. Thus *C. elegans*, given the genetic and molecular tools available, may be the place to look for genes that interact with known or postulated intracellular signalling cascades, as regulators or effectors.

When initiating a project, it is important to first assess the level of conservation expected between the gene of interest and the hoped-for *C. elegans* homologue. The best-guess date for the splitting of the nematode, chordate, and arthropod lineages is currently in the order of 750 million years (38, 39). The branching order of the major metazoan lineages is still unclear (40). There are conflicting models based on morphological, embryological, and molecular data (41–43). The commonly accepted hypothesis is that nematodes, as pseudocoelomate triploblasts, are basal to both the arthropods and chordates (and other coelomate triploblasts). Under this model, a gene present in flies and mice need not be present in nematodes, but if it is it is likely to show even lower levels of similarity than the two 'higher' organism genes do to each other. The alternative model proposes that nematodes are more closely related to flies than mice (44), and predicts that a nematode gene would be closer to the fly than the mouse homologue. Extensive analysis of a number of orthologous genes from humans, flies, nematodes, and yeast has thus far failed to resolve this problem (45), but slower-evolving genes tend to support an arthropod–nematode clade.

Different genes evolve with very different modes and tempos (46, 47). A gene with a core role in transcription, DNA superstructure, or the cytoskeleton (such as elongation factors, histones, and tubulins) will be strongly conserved because it interacts with other conserved macromolecular structures (DNA structure, other cytoskeletal elements). Genes that have core enzymatic roles may also show strong conservation if the enzymatic structure is at some conserved optimum for its role. However, genes with roles in exotic functions, or in functions that engage with the dynamics of the external environment (such as those involved in sex determination (48), or xenobiotic detoxification (49–51)) may evolve at much greater rates.

For example, the relationships of the four *C. elegans* TGF-β genes to their vertebrate and arthropod homologues are illustrated in *Figure 1*. Two of the *C. elegans* TGF-βs (encoded by *dbl-1* and *tig-2*) are most like the bone morphogenetic proteins of vertebrates and *decapentaplegic* of *Drosophila*, while the other two (encoded by *daf-7* and *tig-1*) are apparently diverged extensively. It could be hypothesized that *dbl-1* and *tig-2* are involved in a morphogenetic process like their homologues. In contrast, *daf-7* is involved in controlling developmental fates following sensing of internal and external environments, and does not appear to have a clear vertebrate homologue. Other gene families in *C. elegans* are, or have features, peculiar to this nematode (or to nematodes). For example, the *C. elegans* tyrosinase genes form a separate radiation from other tyrosinases, suggesting evolution to novel function (see *Figure 2*).

For any particular gene, the level of similarity to a putative *C. elegans* homologue will depend on:

(a) the phyletic separation of the originating species from *C. elegans*;

(b) the rate of fixation of sequence change in the gene, dependent on the rate of molecular evolution of the species and the level of integration of the function of the protein in organismal biology (thus, transcription factors and ribosomal proteins evolve slowly, while sex determination genes evolve quickly); and

(c) the relative conservation of the particular biochemical pathway between *C. elegans* and the starting organism (an enzyme may have changed in all but a few active site residues if its substrate and modulators are different).

6.1 Finding your homologue

Thus, for any particular project intended to identify the *C. elegans* homologue(s) of a gene of interest, the following approach could be taken.

6.1.1 Search the *C. elegans* genome databases with your gene of interest

Using either the integrated public databases or the *C. elegans* specific databases (see *Box 2* for World Wide Web URLs) perform a BLAST or FASTA search. If you have peptide (or predicted peptide) data, a BLASTP or TBLASTN search should be performed against the relevant databases.

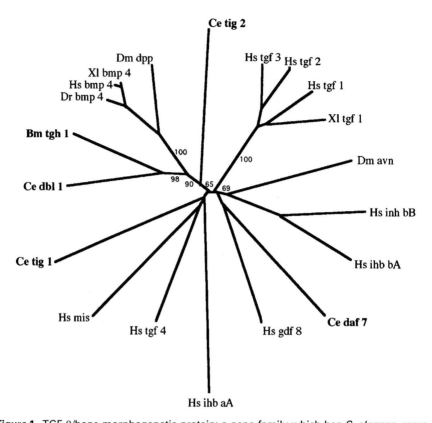

Figure 1. TGF-β/bone morphogenetic protein: a gene family which has *C. elegans* representatives which fall into several classes also found in other model organisms. The *C. elegans* database was searched for predicted proteins similar to the TGF-β family of morphogens (77). The four sequences found were checked in ACeDB (and some adjustment made to the prediction for T25F10.2) and the active peptide domain (TGFs are synthesized as pre-profactors) aligned with representative TGF family members from vertebrates and arthropods. The alignment was used to derive a phylogram using the neighbour-joining algorithm. To test the robustness of the derived tree, bootstrap analysis was used. When significant (> 65%) bootstrap support was found for the deeper branches in the phylogram, the level of support is given. The *C. elegans* TGF-β homologues are as distinct from each other as the different classes of TGFs from other organisms, but two (encoded by *tig-2* and *dbl-1*) are members of the bone morphogenetic protein/decapentaplegic subfamily (78).

Key to sequence labels: Ce daf 7, *C. elegans daf-7* (Wormpep B0412.2); Ce dbl 1, *C. elegans dbl-i* (T25F10.2); Ce tig 2, *C. elegans tig-2* (F39G3.8); Ce tig 1, *C. elegans tig-1* (C53D6.2); Bm tgh 1, *Brugia malayi tgh-1*; Hs tgf 1, Human *transforming growth factor-beta-1*; Hs tgf 2, Human *tgf-beta-2*; Hs tgf 3, Human *tgf-beta-3*; Hs tgf 4, Human *tgf-beta-4*; Xl tgf 1, Xenopus *tgf-beta-1*; Hs bmp 4, Human *bone morphogenetic protein 4*; Dr bmp 4, Danio (zebrafish) *bone morphogenetic protein 4*; Xl bmp 4, Xenopus *bone morphogenetic protein 4*; Dm dpp, *Drosophila decapentaplegic*; Hs ihb aA, Human *inhibin alpha chain*; Hs ihb bA, Human *inhibin beta A chain*; Hs inh bB, Human *inhibin beta B chain*; Dm avn, *Drosophila activin*; Hs mis, Human *Mullerian inhibiting substance*; Hs gdf 8, Human myostatin (*growth and differentiation factor 8*).

Note added in press: *tig-2* has now been identified as *unc-129*.

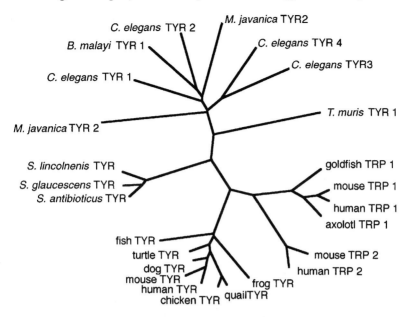

Figure 2. Tyrosinase genes of nematodes form a distinct subfamily, perhaps indicating novel function. *C. elegans* has four genes encoding proteins similar to tyrosinases, enzymes involved in phenoloxidase reactions. In humans and other vertebrates these enzymes are part of the melanin biosynthesis pathway: their biological role in nematodes is unknown. The active-site domains of the four *C. elegans* TYR proteins were aligned to protein sequences from three other nematodes (the vertebrate parasites *Trichuris muris* and *Brugia malayi* and the plant parasite *Meloidogyne javanica*; D. Gerrits and M. Blaxter, unpublished observations), and to tyrosinases from bacteria (*Streptomyces* spp.) (79) and vertebrates. Phylogenetic analysis was performed as for *Figure 1*. The vertebrate proteins include two major families, the true tyrosinases and the tyrosinase-related protein (TRP), which have related enzymatic function (80). The nematode genes define a radiation distinct from the other tyrosinase gene lineages.

Perform a basic set of BLAST (52–54) searches against the *C. elegans* genomic, EST, and protein-sequence databases. TBLASTN and BLASTP searches, which identify regions of similarity between a peptide sequence and either a six-frame translation of a nucleotide database or a peptide database, will be most informative. If you have a nucleotide sequence, the most sensitive search is made using BLASTX against a protein database, to detect potential homologues of peptides encoded by your gene. Unless the search sequence is from a closely related nematode, a nucleotide search (BLASTN) is unlikely to be informative.

6.1.2 Align the *C. elegans* and other homologues

If your search is successful, in that you have identified a potential *C. elegans* homologue with a similarity score significantly above that expected by chance, an informative next step is to align this *C. elegans* gene with your starting gene and any other homologues known from other organisms. Are residues known or predicted to be important in gene function conserved? Is there a shared pattern of conservation of residues between all the potential homologues? One pattern often found is that the *C. elegans* gene shares a domain with genes from other species, but this domain is in a different environment from the corresponding domain in the search sequence. For example, there is a large family of epidermal growth factor-like, domain-containing genes in *C. elegans* (approximately 110 genes (5)) and searching with EGF domain-containing query sequences will probably identify many of these. If you know or suspect that your protein contains one of these common domains, it can be removed or masked before performing the search. You can detect domains by using one of the protein-motif search systems described below (see Section 6.1.3). There are, of course, cases where it is the juxtaposition of such conserved modules that defines the function of a protein.

6.1.3 When no clear homologue is found

If no clear homologue is found, it may be that the *C. elegans* gene has diverged so far from your search sequence that simple algorithms cannot detect similarity. For example, the *C. elegans* globin is a haem-containing, oxygen-binding globin with the same predicted tertiary structure as myoglobin, but it is only 19% identical to vertebrate globins (55, 56). More complex search strategies can be used which rely on the conserved features of a set of aligned protein sequences. These approaches, matrix searching and motif searching, require that you have more than one sequence of your gene of interest.

Align all the known homologues of the gene of interest. Use the alignment to identify highly conserved regions. If there are multiple different subfamilies of the gene, identify defining characteristics of the subfamily required. For example, in homeobox genes, the homeobox flanking regions may be more informative of orthology than the homeobox itself. Conversely, some segments of a protein may be under much less constraint than others, and will thus be less likely to be conserved across species or phyla.

Prealigned and analysed protein families are available in two databases: Pfam (57) and BLOCKS (58, 59). These databases have been constructed from the public protein-sequence databases by recursive cross-comparison of all the sequences, and aim to identify conserved sequence signatures in proteins or subdomains of proteins. Both databases are available and searchable via the World Wide Web (see *Box 2*). Importantly, the protein sequences from the *C. elegans* genome have been incorporated into these databases, so a direct search of these databases with your query sequence may serve to identify the *C. elegans* homologue without the need for additional searching.

6.1.4 Gene families: orthology and paralogy

Genes are commonly duplicated in evolution, and the different copies can then diverge in function. It is important when comparing genes belonging to gene families that like is compared to like. For example, in mammals, myoglobin, α-globin, and β-globin are all homologues of each other, but in comparing gene function between mammals it is important that myoglobin from species A is compared to myoglobin from species B, and not to the other globins, which arose by gene duplication 4–500 million years ago. Gene pairs that arise by duplication are called paralogues, and those arising by organismal speciation are termed orthologues. Orthologous comparisons are more informative than paralogous ones.

For gene families, it is thus important to perform some analysis of molecular (or gene) phylogeny to address the question of what sort of features might distinguish an orthologue from a paralogue. If a set of orthologous sequences is available, a phylogeny can also be used to estimate the level of divergence over unit time. If the known genes are all from mammals, a mammalian phylogeny can be used to identify the expected rate of substitution.

6.1.5 Profile and hidden Markov model searching

If the protein of interest is part of a family in Pfam or BLOCKS, then those databases can be queried to search for *C. elegans* members. Alternatively, a set of aligned sequences can be used to create an *ad-hoc* profile, which can be used to search protein databases (60). These profile-search algorithms are integrated into the Wisconsin Genetics Center Group (GCG) suite of analysis programs. An advance on simple profile searching is the use of hidden Markov models (HMM) to query sequence databases (61). The Pfam databases can be queried with HMM, and the HMM building and searching software is available for local installation (see *Box 2*).

6.1.6 Verification

If potential homologues are identified, retrieve them from the public databases and perform a multiple sequence alignment. Are the expected conserved residues retained in the *C. elegans* homologue? If not, are replacements conservative or not? One of the commonest findings is that, although only one copy of the gene-of-interest is known from the starting organism, analysis of the *C. elegans* genome reveals several homologues (see, for example, *Figure 2*). This may be indicative of the existence of a similar gene family in the starting organisms, as yet unidentified. Additional sequence and biological clues may have to be used to decide which is the putative orthologue and which may therefore mediate similar biological function. Different members of a gene family may have distinct yet overlapping function, and this should be borne in mind when designing knockout experiments.

The *C. elegans* genome sequence is being extensively annotated by the genome centres. A proportion of this annotation is parsed into the public sequence databases, but there is a significant extra set of information available in the genome database ACeDB. In addition, it is possible to search

ACeDB with text queries, and thus find all the *C. elegans* genes with noted similarity to serine–threonine kinase genes, or *ras* genes, etc. ACeDB also displays complementary information, such as the existence of ESTs covering the gene, the presence of linked transposon insertions, and the structure of the genetic map in the area of the gene. Thus an important initial component of any project involving the identification of a gene and investigation of its function should be an examination of the data present in ACeDB (see Chapter 3).

7. Caveats on genome sequence data and gene predictions

The *C. elegans* genome sequence is determined on both strands to a density of between six and eight reads per base (3). The expected number of errors is thus very low. Differences between the 'canonical' genome sequence and the sequence present in a laboratory strain of *C. elegans* can arise from mutation in the nematode genome, or mutation in the cosmid or YAC sequence, or technical sequence error. The genome consortium is always pleased to hear of sequence differences between the database and locally determined data. The commonest problems are with small genes or exons of genes, which may be missed, and in the correct prediction of the 5′ and 3′ ends of genes. The 5′ untranslated regions of nematode genes tend to be very short. The N-terminus of proteins also tends to be less conserved than more central regions. Many *C. elegans* genes are *trans*-spliced to a short leader exon, and the 3′ splice-site acceptor consensus for this event is the same as that for conventional *cis* splicing (27, 62, 63). These circumstances conspire to make it difficult to predict small upstream exons. At the 3′ end, a similar problem is found: it is not yet possible to predict the 3′ untranslated portion of genes accurately, especially if they contain introns.

8. Other nematode genome projects

The *C. elegans* project has inspired a genomics-based approach for several other nematodes. Primary among these is *C. briggsae*, a closely related rhabditid, which can be cultured as for *C. elegans* and shares most of the basic biology. *C. briggsae* has been used for many years as a comparator for *C. elegans*, and this has been especially true in the area of comparative sequence analysis. The Washington University Genome Center (St. Louis, USA) has initiated a limited genome sequencing project on *C. briggsae*, based on a physical map of fosmid clones contiguated by fingerprinting. A number of ESTs (2400) have also been generated. The genome sequencing project has so far yielded over 5 Mb, which reveals extensive synteny with *C. elegans*. Gene order and orientation is, in general, conserved. The major differences detected thus far are: (a) the presence/absence of transposon-like elements;

and (b) *C. briggsae* intergenic regions and introns tend to be shorter than the corresponding *C. elegans* ones. Comparison of genes between the two species has two benefits. The high degree of coding sequence conservation serves to confirm gene predictions made from raw genomic sequence. Comparison of genes between the species can reveal patterns of conservation and divergence that are informative for structure–function studies (48). Outside the coding regions, the phylogenetic distance between the two species is such that most intergenic and intron sequence is significantly diverged. Within this diverged sequence islands of high levels of conservation can identify promoter and enhancer elements: as the protein transcription factors involved in gene regulation are conserved, their binding sites are likewise maintained (64–66). In cases where evolution of sequence is rapid, such as in the sex determination genes, the conservation of synteny has permitted the cloning of homologues too diverged to detect by hybridization methods (48).

The nematodes are an ancient and diverse phylum (67, 68). Many species are parasites of humans, animals, and plants, and one of the promising areas of application of the *C. elegans* genome initiative is to the analysis of parasites (41, 69). While the extreme phylogenetic distances can make it problematic to transfer directly to a parasitic taxon from *C. elegans* (for example, to use a *C. elegans* gene to isolate the parasite's homologue), a parallel genomics approach to other nematode species can yield information which permits transfer. Genetic maps are being built for the free-living *Pristionchus pacificus* (70–72) and important parasites (73). *P. pacificus* is being developed as a second model nematode for use in analysis of the evolution of developmental processes (70–72). For these, and several others, EST datasets have been generated for the purposes of gene discovery and comparison to *C. elegans*. The largest non-*C. elegans* nematode sequence dataset is that for the filarial nematode *Brugia malayi*, a parasite of humans (74, 75). Smaller EST datasets are available in the public databases for other filaria (*Onchocerca volvulus, Wuchereria bancrofti,* and *Loa loa*), additional gut parasites (*Strongyloides stercoralis* (76), *Toxocara canis, Necator americanus, Ascaris suum,* and *Trichuris muris*) and the free-living *P. pacificus*. ESTs are being generated for additional nematode species, including plant parasites (*Meloidogyne* spp.). These datasets provide independent confirmation of the correct prediction of genes from the genomic DNA, and highlight conservation of coding potential. In many cases, the comparison can be more informative than a *C. elegans—C. briggsae* one, as the amount of change will be significantly greater.

References

1. Sulston, J. E. and Brenner, S. (1974). *Genetics*, **77**, 95.
2. Waterston, R., Sulston, J. E., and Coulson, A. R. (1997). In C. elegans *II* (ed. D. Riddle, T. Blumenthal, B. Meyer, and J. Priess), p. 23. Cold Spring Harbor Laboratory Press, Cold Spring Harbor, NY.

3. The *C. elegans* Genome Sequencing Consortium (1998). *Science*, **282**, 2012.
4. Hodgkin, J., Plasterk, R. H. A., and Waterston, R. H. (1995). *Science*, **270**, 410.
5. Hutter, H., personal communication.
6. Clark, N. D. and Berg, J. M. (1998). *Science*, **282**, 2018.
7. Ruvkun, G. and Hobert, O. (1998). *Science*, **282**, 2033.
8. Bargman, C. I. (1998). *Science*, **282**, 2028.
9. Kuwabara, P. E. (1997). Trends Genet., **13**, 455.
10. Coulson, A. R., Sulston, J. E., Brenner, S., and Karn, J. (1986). *Proc. Natl Acad. Sci. USA*, **83**, 7821.
11. Coulson, A., Huynh, C., Kozono, Y., and Shownkeen, R. (1996). In Caenorhabditis elegans. *Modern biological analysis of an organism* (ed. H. F. Epstein, and D. C. Shakes), p. 533. Academic Press, San Diego, CA.
12. Coulson, A., Waterston, R., Kiff, J., Sulston, J., and Kohara, Y. (1988). *Nature*, **335**, 184.
13. Sulston, J., Du, Z., Thomas, K., Wilson, R., Hillier, L., Staden, R., Halloran, N., Green, P., Thierry-Mieg, J., Qiu, L., Dear, S., Coulson, A., Craxton, M., Durbin, R., Berks, M., Metzstein, M., Hawkins, T., Ainscough, R., and Waterston, R. (1992). *Nature*, **356**, 37.
14. Wilson, R., Ainscough, R., Anderson, K., Baynes, C., Berks, M., Bonfield, J., Burton, J., Connell, M., Copsey, T., Cooper, J., Coulson, A., Craxton, M., Dear, S., Du, Z., Durbin, R., Favello, A., Fraser, A., Fulton, L., Gardner, A., Green, P., Hawkins, T., Hillier, L., Jier, M., Johnston, L., Jones, M., Kershaw, J., Kirsten, J., Laisster, N., Latrielle, P., Lightning, J., Lloyd, C., Mortimore, B., O'Callaghan, M., Parsons, J., Percy, C., Rifken, L., Roopra, A., Saunders, D., Shownkeen, R., Sims, M., Smaldon, N., Smith, A., Smith, M., Sonnhammer, E., Staden, R., Sulston, J., Thierry-Mieg, J., Thomas, K., Vaudin, M., Vaughan, K., Waterston, R., Watson, A., Weinstock, L., Wilkinson-Sproat, J., and Wohldman, P. (1994). *Nature*, **368**, 32.
15. Thierry-Mieg, J. and Durbin, R. (1992). *Cahiers IMABIO*, **5**, 15.
16. Durbin, R. and Thierry-Mieg, J. (1994). In *Computational methods in genome research* pp. 45–55 (ed. S. Suhai). Plenum Press, New York.
17. Dunham, I., Durbin, R., Thierry-Mieg, J., and Bentley, D. (1994). In *Guide to human genome computing* (ed. M. Bishop), p. 111. Academic Press, San Diego, CA.
18. Eeckman, F. H. and Durbin, R. (1996). In Caenorhabditis elegans. *Modern biological analysis of an organism* (ed. H. F. Epstein and D. C. Shakes), p. 583. Academic Press, San Diego, CA.
19. Ivens, A. C. and Little, P. F. R. (1995). In *DNA Cloning: A Practical Approach III.* (ed. D. M. Glover and B. D. Hames), p. 1. IRL Press, Oxford.
20. Brenner, S. (1974). *Genetics*, **77**, 71.
21. McCombie, W. R., Adams, M. D., Kelley, J. M., FitzGerald, M. G., Utterback, T. R., Khan, M., Dubnick, M., Kerlavage, A. R., Venter, J. C., and Fields, C. (1992). *Nature Genet.*, **1**, 124.
22. Waterston, R., Martin, C., Craxton, M., Huynh, C., Coulson, A., Hillier, L., Durbin, R., Green, P., Shownkeen, R., Halloran, N., Metzstein, M., Hawkins, T., Wilson, R., Berks, M., Du, Z., Thomas, K., Thierry-Mieg, J., and Sulston, J. (1992). *Nature Genet.*, **1**, 114.
23. Fulton, L. L., Hillier, L., and Wilson, R. K. (1996). In Caenorhabditis elegans.

Modern biological analysis of an organism (ed. H. F. Epstein and D. C. Shakes), p. 571. Academic Press, San Diego, CA.

24. Kohara, Y. (1996). *Tanpakushitsu Kakusan Koso*, **41**, 715.
25. Favello, A., Hillier, L., and Wilson, R. K. (1996). In Caenorhabditis elegans. *Modern biological analysis of an organism* (ed. H. F. Epstein and D. C. Shakes), p. 551. Academic Press, San Diego, CA.
26. Fields, C. (1990). *Nucleic Acids Res.*, **18**, 1509.
27. Blumenthal, T. and Steward, K. (1997). In C. elegans *II* (ed. D. Riddle, T. Blumenthal, B. Meyer, and J. Priess), p. 117. Cold Spring Harbor Laboratory Press, Cold Spring Harbor, NY.
28. Nacleiro, G., Cangiano, G., Coulson, A., Levitt, A., Ruvolo, V., and La Volpe, A. (1992). *J. Mol. Biol.*, **226**, 159.
29. La Volpe, A. (1994). *J. Mol. Evol.*, **39**, 473.
30. Cangiano, G. and La Volpe, A. (1993). *Nucleic Acids Res.*, **21**, 1133.
31. Herman, R. K. and Shaw, J. E. (1987). *Trends Genet.*, **3**, 222.
32. Plasterk, R. H. A. and van Leunen, H. G. A. M. (1997). In C. elegans *II* (ed. D. Riddle, T. Blumenthal, B. Meyer, and J. Priess), p. 97. Cold Spring Harbor Laboratory Press, Cold Spring Harbor, NY.
33. Janke, D. L., Schein, J. E., Ha, T., Franz, N. W., O'Neil, N. J., Vatcher, G. P., Stewart, H. I., Kuervers, L. M., Baillie, D. L., and Rose, A. M. (1997). *Genome Res.*, **10**, 974.
34. Korswagen, H. C., Durbin, R. M., Smits, M. T., and Plasterk, R. H. A. (1996). *Proc. Natl Acad. Sci. USA*, **93**, 14680.
35. Kimura, K. D., Tissenbaum, H. A., Liu, Y., and Ruvkun, G. (1997). *Science*, **277**, 942.
36. Estevez, M., Attisano, L., Wrana, J. L., Albert, P. S., Massagué, J., and Riddle, D. L. (1993). *Nature*, **365**, 644.
37. Morris, J. Z., Tissenbaum, H. A., and Ruvkun, G. (1996). *Nature*, **382**, 536.
38. Bowring, S. A., Grotzinger, J. P., Isachsen, C. E., Knoll, A. H., Pelechaty, S. M., and Kosolov, P. (1993). *Science*, **261**, 1293.
39. Conway Morris, S. (1993). *Nature*, **361**, 219.
40. Nielsen, C. (1995). *Animal evolution. Interrelationships of the living phyla*, p. 467. Oxford University Press, Oxford.
41. Blaxter, M. L. (1998). *Science*, **282**, 2041.
42. Fitch, D. H. A. and Thomas, W. K. (1997). In C. elegans *II* (ed. D. Riddle, T. Blumenthal, B. Meyer, and J. Priess), p. 815. Cold Spring Harbor Laboratory Press, Cold Spring Harbor, New York.
43. Baldwin, J. G., Frisse, L. M., Vida, J. T., Eddleman, C. D., and Thomas, W. K. (1997). *Mol. Phylogenet. Evol.*, **8**, 249.
44. Aguinaldo, A. M. A., Turbeville, J. M., Linford, L. S., Rivera, M. C., Garey, J. R., Raff, R. A., and Lake, J. A. (1997). *Nature*, **387**, 489.
45. Mushegian, A. R., Garey, J. R., Martin, J., and Liu, L. X. (1998). *Genome Res.*, **8**, 590.
46. Thomas, W. K. and Wilson, A. C. (1991). *Genetics*, **128**, 269.
47. Li, W.-H. and Graur, D. (1991). *Fundamentals of molecular evolution*. Sinauer Associates, Sunderland, MA.
48. Kuwabara, P. and Shah, S. (1994). *Nucleic Acids Res.*, **22**, 159.
49. Lincke, C. R., Broeks, A., The, I., Plasterk, R. H. A., and Borst, P. (1993). *EMBO J.*, **12**, 1615.

50. Broeks, A., Janssen, H. R. W. M., Calafat, J., and Plasterk, R. H. A. (1995). *EMBO J.*, **14**, 1858.
51. Broeks, A., Gerrard, B., Allikmets, R., Dean, M., and Plasterk, R. H. A. (1996). *EMBO J.*, **15**, 6132.
52. Altschul, S. F., Gish, W., Miller, W., Myers, E. W., and Lipman, D. J. (1990). *J. Mol. Biol.*, **215**, 403.
53. States, D. J., Gish, W., and Altschul, S. F. (1991). *Methods*, **3**, 66.
54. Altschul, S. F. (1993). *J. Mol. Evol.*, **36**, 290.
55. Kloek, A. P., Sherman, D. R., and Goldberg, D. E. (1993). *Gene*, **129**, 215.
56. Blaxter, M. L. (1993). *Parasitol. Today*, **9**, 353.
57. Sonnehammer, E. L. L. and Kahn, D. (1994). *Protein Sci.*, **3**, 482.
58. Henikoff, S. and Henikoff, J. G. (1992). *Proc. Natl Acad. Sci. USA*, **89**, 10915.
59. Tatusov, R. L., Altschul, S. F., and Koonin, E. V. (1994). *Proc. Natl Acad. Sci. USA*, **91**, 12091.
60. Gribskov, M., McLachlan, A. D., and Eisenberg, D. (1987). *Proc. Natl Acad. Sci. USA*, **84**, 4355.
61. Durbin, R., Eddy, S., Krogh, A., and Mitchison, G. (1998). *Biological sequence analysis: probabilistic models of proteins and nucleic acids. A tutorial introduction to hidden Markov models and other probabilistic modelling approaches in computational sequence analysis*, p. 356. Cambridge University Press, Cambridge.
62. Speith, J., Brooke, G., Kuersten, S., Lea, K., and Blumenthal, T. (1993). *Cell*, **73**, 521.
63. Blaxter, M. L. and Liu, L. X. (1996). *Int. J. Parasitol.*, **26**, 1025.
64. Heschl, M. F. P. and Baillie, D. L. (1990). *J. Mol. Evol.*, **31**, 3.
65. Heschl, M. F. P. and Baillie, D. L. (1989). *DNA*, **8**, 233.
66. Gilleard, J. S., Barry, J. D., and Johnstone, I. L. (1997). *Mol. Cell. Biol.*, **17**, 2301.
67. Conway Morris, S. (1981). *Parasitology*, **82**, 489.
68. Blaxter, M. L., De Ley, P., Garey, J., Liu, L. X., Scheldeman, P., Vierstraete, A., Vanfleteren, J., Mackey, L. Y., Dorris, M., Frisse, L. M., Vida, J. T., and Thomas, W. K. (1998). *Nature*, **392**, 71.
69. Blaxter, M. L. and Bird, D. M. (1997). In C. elegans II (ed. D. Riddle, T. Blumenthal, B. Meyer, and J. Priess), p. 851. Cold Spring Harbor Laboratory Press, Cold Spring Harbor, NY.
70. Sommer, R. J., Carta, L. K., Kim, S.-Y., and Sternberg, P. W. (1996). *Fundam. Appl. Nematol.*, **19**, 511.
71. Schlak, I., Eizinger, A., and Sommer, R. (1997). *J. Zool. Syst. Evol. Res.*, **35**, 137.
72. Sommer, R. J. and Sternberg, P. W. (1996). *Curr. Biol.*, **6**, 52.
73. Jeroen, N. A. M., van der Voort, R., Roosien, J., van Zandvoort, P. M., Folkertsma, R. T., van Enckevort, E. L. J. G., Janssen, R., Gommers, F. J., and Bakker, J. (1994). In *NATO ARW: advances in molecular plant nematology* (ed. F. Lamberti, C. De Giorgi, and D. M. Bird), p. 57. Plenum Press, New York.
74. Blaxter, M. L. (1995). *Parasitol. Today*, **11**, 811.
75. Blaxter, M. L., Raghavan, N., Ghosh, I., Guiliano, D., Lu, W., Williams, S. A., Slatko, B., and Scott, A. L. (1996). *Mol. Biochem. Parasitol.*, **77**, 77.
76. Moore, T. A., Ramachandran, S., Gam, A. A., Neva, F. A., Lu, W., Saunders, L., Williams, S. A., and Nutman, T. B. (1996). *Mol. Biochem. Parasitol.*, **79**, 243.
77. Ren, P., Lin, C.-S., Johnsen, R., Albert, P. S., Pilgrim, D., and Riddle, D. L. (1996). *Science*, **274**, 1389.

78. Sekelsky, J. J., Newfeld, S. J., Raftery, L. A., Chartoff, E. H., and Gelbart, W. M. (1995). *Genetics*, **139**, 1347.
79. Jackman, M. P., Hanjal, A., and Lerch, K. (1991). *Biochem. J.*, **274**, 707.
80. Tsukamoto, K., Jackson, I. J., Urabe, K., Montague, P. M., and Hearing, V. J. (1992). *EMBO J.*, **11**, 519.

3

C. elegans and the Web

JEAN THIERRY-MIEG, DANIELLE THIERRY-MIEG,
MICHEL POTDEVIN, and LINCOLN STEIN

1. Introduction

With the completion of the *C. elegans* sequencing project, it is now a frequent event for a biologist working on another species to detect a homology to a nematode sequence during a routine BLAST search. Starting from such a link, or from a text search, we show, step by step, how a scientist, who does not have specialized knowledge of *C. elegans* biology, may use the Web to explore the *C. elegans* data. We show how to access the genomic DNA and cDNA sequences, look at the graphic displays of the sequence and the genetic map, and search for mutant phenotypes. Finally, we explain how to contact the relevant *C. elegans* laboratories to ask for further help, information, and possibly strains, clones, or other material from which may develop a new and fruitful collaboration.

Research on *C. elegans* has been built on co-operation, with the belief that this is the most efficient form for scientific endeavour. The free exchange of data has led to the central accumulation of a large body of data on all aspects of *C. elegans* biology (1). These results have now been capped by the efforts of the *C. elegans* sequencing consortium, led by John Sulston and Bob Waterston, which has now almost entirely established, with great accuracy, the DNA sequence of the reference strain (2). The study by Yuji Kohara of over 100 000 expressed sequence tags (ESTs) (3), in conjunction with the genomic sequence, gives the exact splicing of over half of the expressed genes. Previously, *C. elegans* data were interesting mainly to *C. elegans* specialists, but the genomic sequence is now providing links to biologists working with other organisms.

Information about *C. elegans* is stored in the database ACeDB (4). Central to ACeDB are three different windows on the *C. elegans* genome. The Sequence window sees the genome as a string of nucleotide bases, with annotation arising from interpretation of those data. The Physical map window sees the genome as a set of genomic DNA clones covering the genome, the starting point for the sequencing project. The Genetic window sees the genes as detected by mutation and positioned using genetic recombination distances or other genetic mapping techniques. ACeDB includes all the known interconnections between these different views on the genome as well as connections to other types of biological data or additional information.

Up to now, ACeDB was used as a tool to support the sequencing project. However, copies were installed in most dedicated *C. elegans* laboratories to support research on other aspects of *C. elegans* biology. These copies require maintenance with frequent updates. To facilitate access to ACeDB for other biologists, we are starting to modify the presentation and have recently installed a set of Web pages on our server at `http://alpha.crbm.cnrs-mop.fr/` using the new AcePerl library (5). These pages provide browsable access to the *C. elegans* data through several general entry points:

- via a BLAST search against the *C. elegans* sequence;
- via a text search over the entire database;
- via gene, sequence, or clone names;
- via the genetic, sequence, or clone maps;
- via author name or bibliography.

The aim of this chapter is to guide users through the nematode data from these starting points until they can establish an appropriate collaboration with a *C. elegans* laboratory.

2. Starting from a sequence homology

Frequently, the first reference to *C. elegans* is found in the output of a sequence homology search of a general database. For example, a BLAST search against the SWISSPROT database on the Web server `http://www.ncbi.nlm.nih.gov` may have revealed a highly significant match to the following entry:

```
>sp|P34373|YLJOCAEEL HYPOTHETICAL 44.3 kDa PROTEIN C50C3.10 IN
CHROMOSOME III
```

A little exploration in SWISSPROT reveals that this entry corresponds to a predicted gene in the *C. elegans* genomic sequence. 'C50C3' is the name of the source cosmid, and 'C50C3.10', which also appears in the SWISSPROT description line, indicates that this is predicted to be gene number 10 in this cosmid. Aside from this, there is little additional information in SWISSPROT that would allow us to evaluate the biological significance of this match or to make contacts with potential collaborators within the *C. elegans* community.

If we now connect to the ACeDB server `http://alpha.crbm.cnrs-mop.fr/`, go to the browser, select the sequence field, and type the sequence name (C50C3.10) and press the return key in the text entry, a summary appears of all the relevant information known about this sequence (see *Figure 1*). Information on this page includes structural information about the sequence (exon/intron boundaries and so forth), the predicted protein translation, and comments made by annotators. All in all, this is not so very

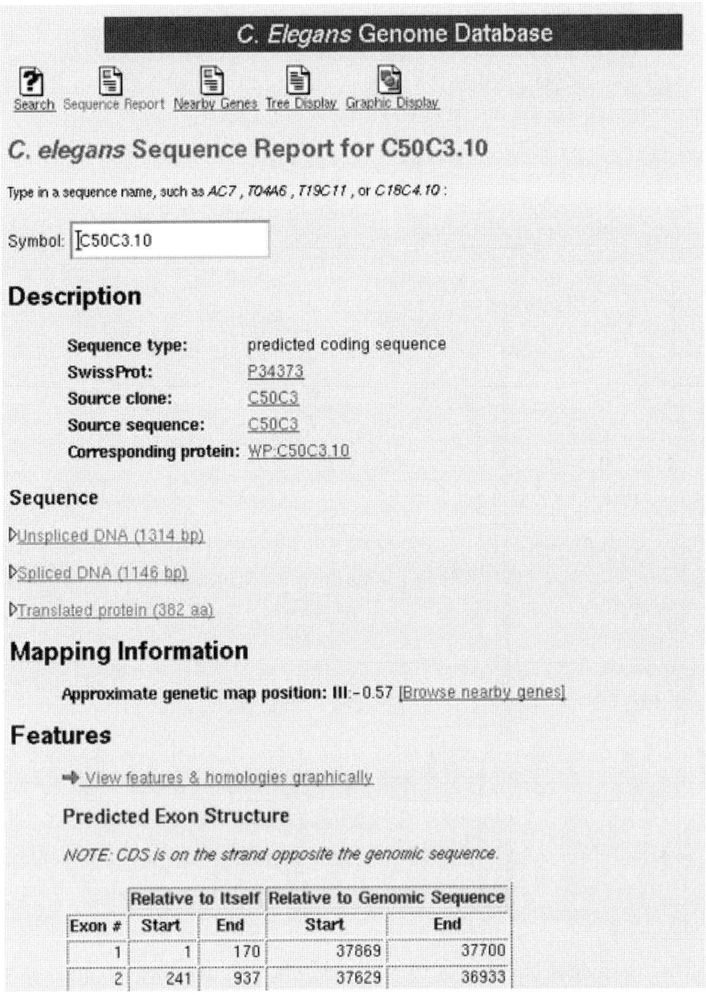

Figure 1. The '*C. elegans* sequence report for C50C3.10' window. This is reached from the browser at `http://alpha.crbm.cnrs-mop.fr/`, as described in the text. (This will appear in colour on the computer screen.)

different from the SWISSPROT entry because, in fact, the SWISSPROT entry was created from the ACeDB entry. However, our browser provides many links directly to the rest of the *C. elegans* data.

2.1 Graphical view of the sequence and of the cDNAs

If we follow the link, at the top of the page in *Figure 1*, marked 'Graphic display' we will see a screen similar to the one shown in *Figure 2*. Selecting the 'Zoom out' button three times to reduce the scale and then clicking on the

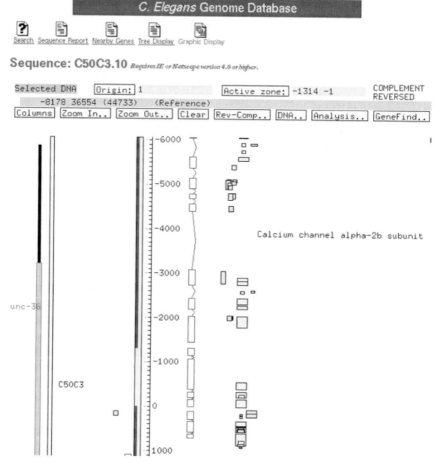

Figure 2. The 'Sequence for C50C3.10' 'Graphic display' window. This is reached from the 'Sequence report' window (*Figure 1*), as described in the text. (This will appear in colour on the computer screen.)

'Rev-Comp' (reverse and complement) button will create exactly the screen shown in *Figure 2*. On this diagram, which is exported directly from the ACeDB database, the chain of connected boxes (blue on the screen), just to the right of the coordinate axis, shows the predicted genes. These genes and their structures correspond to the result of the GENEFINDER prediction program and to the text report discussed previously. Genes to the right of the yellow bar are transcribed down the page and those to the left, up the page. The results of previously scanned BLAST analysis appear as a cloud of boxes (will be shaded light blue on your computer screen) and may also be seen as text tables by using the 'Sequence report' button at the top of the window.

However, the cDNA sequence is also available for about half the genes, providing a reliable determination of gene structure. The cDNAs come mainly from the laboratory of Yuji Kohara. The individual cDNAs are shown as yellow boxes. By the time of publication, it is anticipated that the diagram will be modified so that cDNAs appear as arrows with the reconstructed gene in purple. Our example is, in fact, a single transcribed gene that was mistakenly broken into three predicted genes by the prediction program. The names of the genes, C50C3.9, .10, and .11 in this case, are revealed by clicking on the gene structures and thereby moving to the 'Sequence report' window for each gene in turn. These three predicted genes are close together and transcribed in the same direction, and cDNAs bridging these units reveal that they are, in fact, one gene. EST sequences for opposite ends of a cDNA clone are given the suffix .5 or .3, depending on whether they correspond to the start or end of the cDNA. The 'Sequence report' window for C50C3.11 also reveals that this predicted gene corresponds to the well-studied gene *unc-36*.

Such mistakes in gene structure prediction are rather frequent, because although it is relatively easy to predict exons, it is much harder to predict the exact limits of a gene. The database contains over 100 000 cDNA sequences and many of them have not yet been edited. Editing of particular cDNA sequences of specific interest can be prioritized by sending us an e-mail using the link provided. Editing can provide the exact position of the introns, as well as the start and end of the gene, often correcting a GENEFINDER prediction, and may reveal the existence of alternative splicing. In some cases, gene-expression pattern images, obtained by *in-situ* hybridization using the cDNA clones, have been collected and are available on Kohara's Web site (3). The cDNA clones themselves may be requested from Kohara.

From the 'Graphic sequence' display, more analysis could be performed if working with a local copy of ACeDB (see below), e.g. it is possible to select oligos in a given region to perform a PCR. This is not yet possible over the Web because of various technical limitations and because the network is too slow for this type of interactive application.

2.2 The genetic map

If you are fortunate, the *C. elegans* sequence, for which you have detected a homology, may be found to correspond to a gene defined classically, through mutation. For the moment, this is still rather rare. Indeed, although approximately 1500 classical genes in *C. elegans* have been characterized through a mutant phenotype, only a fraction have already been matched to one of the 19 000 genes predicted from the analysis of the sequence. Nevertheless, because the genome sequence is virtually complete, the position in the sequence information is sufficient to locate, by interpolation, a corresponding region on the genetic map, providing the mutant phenotypes of the known classical genes in that area.

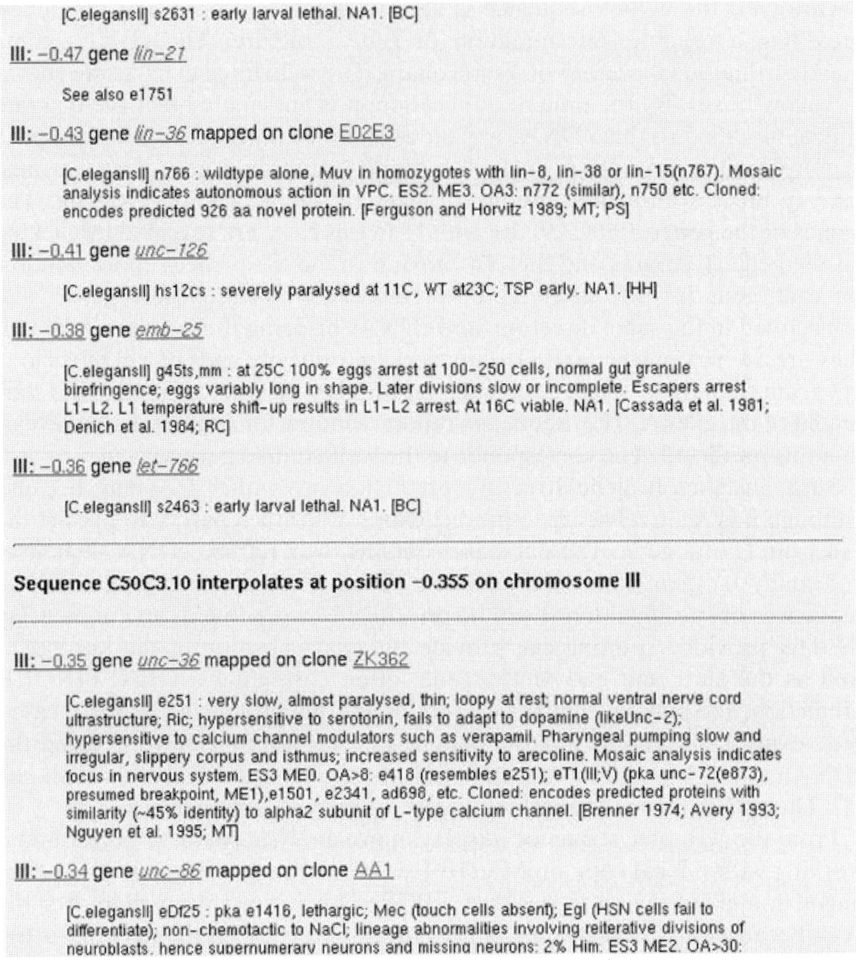

[C.elegansll] s2631 : early larval lethal. NA1. [BC]

III: −0.47 gene *lin−21*

    See also e1751

III: −0.43 gene *lin−36* mapped on clone E02E3

[C.elegansll] n766 : wildtype alone, Muv in homozygotes with lin−8, lin−38 or lin−15(n767). Mosaic analysis indicates autonomous action in VPC. ES2. ME3. OA3: n772 (similar), n750 etc. Cloned: encodes predicted 926 aa novel protein. [Ferguson and Horvitz 1989; MT; PS]

III: −0.41 gene *unc−126*

[C.elegansll] hs12cs : severely paralysed at 11C, WT at23C; TSP early. NA1. [HH]

III: −0.38 gene *emb−25*

[C.elegansll] g45ts,mm : at 25C 100% eggs arrest at 100−250 cells, normal gut granule birefringence; eggs variably long in shape. Later divisions slow or incomplete. Escapers arrest L1−L2. L1 temperature shift−up results in L1−L2 arrest. At 16C viable. NA1. [Cassada et al. 1981; Denich et al. 1984; RC]

III: −0.36 gene *let−766*

[C.elegansll] s2463 : early larval lethal. NA1. [BC]

Sequence C50C3.10 interpolates at position −0.355 on chromosome III

III: −0.35 gene *unc−36* mapped on clone ZK362

[C.elegansll] e251 : very slow, almost paralysed, thin; loopy at rest; normal ventral nerve cord ultrastructure; Ric; hypersensitive to serotonin, fails to adapt to dopamine (likeUnc−2); hypersensitive to calcium channel modulators such as verapamil. Pharyngeal pumping slow and irregular, slippery corpus and isthmus; increased sensitivity to arecoline. Mosaic analysis indicates focus in nervous system. ES3 ME0. OA>8: e418 (resembles e251); eT1(III;V) (pka unc−72(e873), presumed breakpoint, ME1),e1501, e2341, ad698, etc. Cloned: encodes predicted proteins with similarity (~45% identity) to alpha2 subunit of L−type calcium channel. [Brenner 1974; Avery 1993; Nguyen et al. 1995; MT]

III: −0.34 gene *unc−86* mapped on clone AA1

[C.elegansll] eDf25 : pka e1416, lethargic; Mec (touch cells absent); Egl (HSN cells fail to differentiate); non−chemotactic to NaCl; lineage abnormalities involving reiterative divisions of neuroblasts. hence supernumerary neurons and missing neurons; 2% Him. ES3 ME2. OA>30;

Figure 3. The 'Nearby genes' window showing the text version of the genetic map in the vicinity of C50C3.10. This is reached from either of the windows shown in *Figures 1* and *2*, as described in the text. (This will appear in colour on the computer screen.)

The link marked 'Nearby genes', in the 'Graphic display' or the 'Sequence report' windows (see *Figure 2*) for C50C3.10, will retrieve a new page that displays a sorted list of genes known to be in the general region of the genomic DNA sequence under study (see *Figure 3*). In the case of C50C3.10, there are several genes in the general area of interest that have not yet been located in the genomic DNA sequence, including *emb-25* and *unc-36*. Brief details of mutant phenotypes are included alongside the gene names. If a phenotype was related to the known function of the original gene that had started this line of enquiry, then you can select the 'Gene link', taking you to a

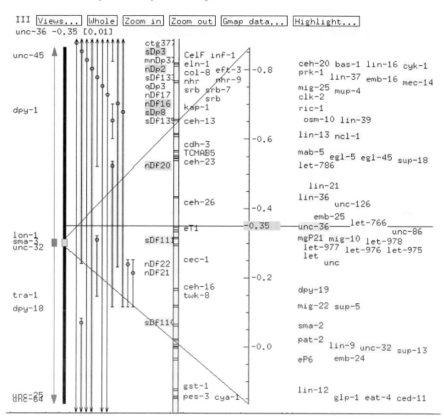

Figure 4. The 'Graphic display' window showing the genetic map in the vicinity of *unc-36*. This is reached from the text version of the genetic map shown in *Figure 3*, as described in the text. (This will appear in colour on the computer screen.)

page, a 'Gene report' window, that describes what is known about this gene in *C. elegans*. From here, you can follow the bibliographic links to papers that describe this gene, and related family members, and retrieve the contact details of the authors.

It is also possible to browse the genetic map in a graphical fashion, either by selecting the link above or below the initially identified position (the chromosome coordinate on the far left of the window as shown in *Figure 3*) or by selecting the 'Graphic display' button at the top of a 'Gene report' window. The genetic map will be re-displayed centred on the new gene. *Figure 4* shows the result when III: –0.35/*unc-36* is the target and after zooming in four times using the zoom button.

The genetic map contains other elements as well, the most important of which are vertical lines (left half of the 'Genetic map' window) indicating chromosomal rearrangements, usually a deficiency (Df, shown as a single line) or a duplication (Dp, shown as a double line), that are available in a characterized worm strain (see Chapter 12, Section 5). Selecting a deficiency with a click of the mouse will cause genes covered by that deficiency to be highlighted. The colours correspond to actual experimental data, not just to the location on the display: green means included in the rearrangement, blue not included, black untested, and red identifies a gene for which there is conflicting data. Just above the blue bar, there is also a button ('Gmap data') that allows you to see the information used to construct the genetic map, e.g. 2-point distances and 3-point ordering information. The same data are accessible as a table using the 'Map data' button at the very top of the display. The interpretation of these data requires some knowledge of *C. elegans* genetics, but can reveal the confidence in a 'Genetic map' position.

3. Starting from a direct BLAST search of the *C. elegans* data

Another possible entry into the *C. elegans* database is via a direct BLAST search of the genomic sequence data. *C. elegans* databases, that can be searched using the BLAST suite of programs, are available on servers at the following locations:

- `http://www.sanger.ac.uk/Projects/C_elegans/blast_server.shtml`
- `http://genome.wustl.edu/gsc/Search/blast.shtml`
- `http://alpha.crbm.cnrs-mop.fr/cgi-bin/ace/blast/elegans`

These servers contains all the sequence finished or in progress from the *C. elegans* consortium. In each case, a positive result will provide the name of the *C. elegans* cosmid or sequence, which you can then follow in the manner described above, to further information about the sequence and genes in the vicinity. On the alpha site described here (the third site listed above), select the BLAST service and paste in your sequence. The result of the BLAST search is presented in the same window and gives directly the name of a *C. elegans* sequence as a direct link to the *C. elegans* database.

For example, using the sequence:

```
gatggccgagaactcgaggactaggatttttgggtgcaatacacagatgtggcctcactggg
aatggctcggaaaaattatgttcacaatttgttttgaaggtagtttattttaaagaataggt
aaaactgtggaaacgaaaaactgcta
```

gives a result which includes the line:

```
>C50C3 [View]
```

Selecting the link labelled 'View' leads to a display that includes a button to the C50C3 sequence report window described in the previous sections.

4. Starting from 'Text search'

If you are pursuing a gene and already have a good idea of the phenotype you expect for a mutation in the homologous gene in *C. elegans*, you may perform a full text search of the *C. elegans* database on the server at `http://alpha.crbm.cnrs-mop.fr/cgi-bin/ace/simple/elegans` and recover the matching objects. Select the 'Text search' page and type, for example, 'fibroblast growth factor' into the search box, not worrying about the capitalization of letters. The system searches the entire database, including paper abstracts, for these words and returns a list of objects, 98 in this example if the in-depth search is selected.

If you are not familiar with *C. elegans* nomenclature, the result may initially appear indecipherable. However, the list is ordered by types and can be easily understood with knowledge of a few key abbreviations. Clones are listed first, then the genes, then papers, then proteins, and so on. In this case, the various GENBANK or SWISSPROT entries correspond to FGF (fibroblast growth factor) related genes with homologies to *C. elegans* DNA sequence. These are very informative, and can lead to the identification of *C. elegans* genes that may be missed in the direct text search because the annotation differed from or lacked the particular phrase searched for. For example 'fibroblast growth factor' fails to recover the gene *let-756* directly, because for some reason it has been labelled 'heparin binding protein' in ACeDB. But *let –756* does show homology to fibroblast growth factor and can be found indirectly (e.g. follow the links through 'C05D11.4' from 'Protein: TR:O15520'). *Let-756* would also be found by simply typing FGF in the 'Text search', which would identify the relevant *Worm breeders' gazette* (*WBG*) articles and Worm meeting abstracts. The names of cDNA clones start with yk if from Yuji Kohara, with cm if from Chris Martin, and with CE if from Craig Venter's library. Finally, the list of bibliographic references includes: regular papers, referenced by the Caenorhabditis Genetics Centre, prefixed with cgc, and unpublished material, such as worm meeting abstracts, prefixed with wm, ecwm, wcwm, mwwm, and ewm, or *WBG* articles, prefixed with wbg. (This unpublished material should not be cited without consent of the authors.) All these names provide direct links into the database.

5. Other types of data

The ACeDB database contains other types of data as well. Everything is accessible from the Web, but for many categories, we have not written a dedicated display. They can be accessed indirectly by navigation or by using the class browser or the ace query forms accessible from the generic 'Simple search' document from where the browser starts. However, most of them are probably not very interesting as entry points. Either they are better accessed indirectly in comprehensive tables, like the raw genetic mapping data, or they are too incomplete to start a search.

In addition to the genes and sequences, the system contains:

- lists of strains, alleles, and raw genetic data, which are accessible from the gene report;
- the physical map, with all the associated data, listing the genomic clones that are available as starting points for molecular studies;
- utilities like restriction enzymes lists and motifs, which are imported from other databases;
- some data on cell lineage and expression patterns, which is, as yet, far from complete;
- bibliographic data, which is accessible indirectly through the 'Text search' interface.

In the future, we hope that the quantity and quality of the data will keep improving and we will develop new displays accordingly.

6. Other ways to look at the ACeDB data

Apart from the explicit http links to GenBank and other external servers, all the graphical data presented in the AcePerl Web interface on our server originate from the ACeDB database. In time more information and links to other servers will be added, in particular to the gene-expression pattern images available from the Yuji Kohara and Ian Hope laboratories, and possibly to results from double-stranded RNA interference or any large-scale database of relevance that becomes available.

Although technically more difficult, it is possible to look directly at ACeDB using your own computer. The interface is faster and more powerful than the Web.

If you have an X-terminal or an X-emulation on your Mac or PC, and if you are not behind a strict 'firewall', you can run ACeDB directly over the Web from our server in Montpellier. Start from our Web page (http://alpha. crbm.cnrs-mop.fr/elegans) and choose X11 display. Because it takes a lot of space on our computer, we only support a few clients at the same time and if the system is busy you may have to retry later. In this way you have access to the complete ACeDB interface, but speed is limited because every action is interpreted on the server side and the answer transmitted over the Net.

Alternatively, the graphic ACeDB client may be downloaded. Unfortunately, so far, this system only works on a Unix machine, or on a PC using the

Linux operating system. However, the code is not so large, and the data will be downloaded automatically when you need them. After a while the system becomes extremely fast and works even if the network fails because the necessary data is cached locally, a bit like the way caching works in Web browsers.

A third alternative is to download the complete ACeDB database, which takes around one gigabyte of disk space. These methods are explained on our Web page (`http://alpha.crbm.cnrs-mop.fr/acedb`).

Finally, if you are able to program in Perl, you can download the complete AcePerl(5) module from our site or from `http://stein.cshl.org/AcePerl`. It is easy to install, fully documented, and is configured by default to access our server containing all the *C. elegans* data. This module provides direct programatic access to the ACeDB database, allowing you to fetch-down relevant data from a script or to create your own Web pages.

7. Conclusions

Out of necessity, this chapter is very sketchy. We are describing a Web site in constant evolution, but which is meant to be self-explanatory. We hope to convey the idea that, on the *C. elegans* Web servers, a very large amount of information is available, over and above the simple result of a BLAST search. It is possible in a few hours to gain a real knowledge of a segment of the *C. elegans* genome and associated biology and to identify relevant *C. elegans* laboratories with which to initiate a collaboration.

Acknowledgements

A Web site is only useful if it contains good data, and we are grateful to all the laboratories who are contributing to the *C. elegans* public database, the big providers like the Sequencing Consortium, but equally all the individuals providing genetic mapping data or molecular identifications. The data are collated by the combined efforts of many people, including Bob Herman, Theresa Stiernagle, Jonathan Hodgkin, Sylvia Martinelli, Yuji Kohara, Tsodas Shini, Richard Durbin, and Ladeana Hillier. The presentation software, with the very many layers that we have combined, Netscape, BLAST, Apache, AcePerl, and ACeDB is the work of very many programmers and we thank them for making it completely free. We are also very grateful to Richard Durbin for our long-standing collaboration and Ian Hope for his numerous suggestions and clarification of this text.

References

1. Riddle, D. L., Blumenthal, T., Meyer, B. J., and Priess, J. R. (ed.) (1997). C. elegans *II.* Cold Spring Harbor Laboratory Press, NY, and Avery L., `http://elegans.swmed.edu`

2. The *C. elegans* Sequencing Consortium (1998). *Science*, **282**, 2012. Sanger
   http://www.sanger.ac.uk/Projects/C_elegans
   Wash-U http://genome. wustl.edu/gsc/C_elegans
3. Kohara, Y. *C. elegans ESTdatabase*, http://www.ddbj.nig.ac.jp/htmls/
   c.elegans/html/CE_INDEX.html
4. Durbin, R. and Thierry-Mieg, J. (1991 onwards). ACeDB documentation code and
   data available from http://alpha.crbm.cnrs-mop.fr/acedb and http://
   www.sanger.ac.uk/software/acedb
5. Stein, L. D. and Thierry-Mieg, J. (1998). *Genome Res.*, **8**, 1308 and
   http://stein.cshl.org/AcePerl

4

Maintenance of *C. elegans*

THERESA STIERNAGLE

1. Introduction

Wild-type and mutant stocks of *Caenorhabditis elegans* are available from the Caenorhabditis Genetics Center (CGC). *C. elegans* is easily grown in the laboratory, with stocks routinely maintained on agar-filled Petri plates. However, they can be grown in liquid culture when larger quantities of worms are needed. Occasional contamination problems are easy to overcome because the eggs are resistant to treatment with bleach. Long-term storage of *C. elegans* in liquid nitrogen is attained through the use of glycerol-containing media. Good rates of recovery from freezing allow stocks to be stored between periods of study without the need for continual transfers.

2. Acquiring strains from the Caenorhabditis Genetics Center (CGC)

2.1 The CGC

The CGC was established at the University of Missouri, Columbia, in 1979. In 1992 the CGC moved from Missouri to the University of Minnesota, St Paul. Funded through a contract with the National Institutes of Health National Center for Research Resources (NIH NCRR), the basic mission of the CGC is to provide *C. elegans* strains and information to scientists initiating or continuing research using *C. elegans*. The CGC strives to acquire, and have available for distribution, stocks representing at least one mutant allele of each published gene and all chromosome rearrangements (deficiencies, duplications, translocations, inversions). In addition, many wild-type isolates of *C. elegans* and several species closely related to *C. elegans* are available. The CGC maintains an up-to-date bibliography of all articles, reviews, and books that discuss *C. elegans*. The *Worm Breeder's Gazette* (WBG), a newsletter published three times per year, is available in hard copy from the CGC for a small 2-year subscription fee, and it is available free of charge on the Internet. The WBG contains a *C. elegans* bibliographical update, announcements and news, a directory of *C. elegans* researchers, and single-page abstracts submitted by *C. elegans* researchers describing current work.

The *C. elegans* genetic nomenclature and the genetic map are maintained by the CGC as a subcontract under the directorship of Jonathan Hodgkin at the MRC Laboratory of Molecular Biology, Cambridge, England. A second subcontract helps Leon Avery (Southwestern Medical Center, University of Texas, Dallas) maintain the *C. elegans* World Wide Web (WWW) server (http://elegans.swmed.edu/) as a source of information about *C. elegans*, including the services of the CGC and descriptions of available strains.

2.2 Requesting strains

Wild-type and over 3500 mutant strains of *C. elegans* are available at no cost to researchers at educational institutions or non-profit organizations. To obtain strains, submit a written request (letter, fax, or e-mail) describing the strains, genes, or alleles desired together with a brief statement of the intended use of the worms. Include within the request the name, complete mailing address, and phone numbers of the person requesting the strain(s).

Researchers at commercial or for-profit organizations are required to pay a fee of $100.00 plus postage for each strain of worms or bacteria ordered. A 'Commercial Sector Order Form' is available from the CGC; a completed form must be received by the CGC before strains will be sent to commercial organizations.

All requests for strains or information should be sent to: Caenorhabditis Genetics Center, University of Minnesota, 250 Biological Sciences Center, 1445 Gortner Avenue, St Paul, MN 55108–1095, USA. Requests can also be faxed (612) 625–5754 or e-mailed stier@biosci.cbs.umn.edu.

2.3 Mailing strains

Upon receipt of a strain request, the CGC will send a response confirming that the order has been received. The requested strains are either thawed from a freezer vial or recovered from a starved plate kept in an incubator at 11 °C. Animals of each strain are placed on freshly prepared Nematode Growth Medium (NGM) agar Petri plates that have been spread with *Escherichia coli* as a food source and allowed to reproduce for several days. The visible phenotype of the stock is noted to confirm that it matches the published description. The Petri plates are then wrapped in Parafilm and mailed to the requester. It takes 7–10 days for a strain request to be received, processed, and filled. If a requested strain has a temperature-sensitive lethal phenotype, the stock is allowed to starve at permissive temperature before the plates are shipped, a step that takes an additional 3–4 days. This results in the formation of heat-resistant dauer larvae (see Chapter 1).

Strains are sent on NGM Petri plates with the worms still feeding on *E. coli* OP50 (or starved, as mentioned above). The plates are shipped in bubble-lined mailing bags via First Class or Air Mail. Postage costs are paid by the

CGC except for commercial orders. Strains can be shipped via Federal Express if the requester provides the CGC with a Federal Express account number to bill the courier costs. Strains sent outside the USA have a United States Postal Service Customs (CN22) sticker applied to the package. The sticker declares that the package contains 'non-pathogenic, non-parasitic biological specimens of no commercial value'. Occasionally the customs departments of other countries require that additional information be applied to the package. In such cases, the requester should supply the CGC with the needed information.

Strains are sent with a 'Strain Information Sheet', which gives the genotype, phenotype, culturing conditions, and derivation of the stock, as well as a bibliographical reference. Requesters are asked to inform the CGC of the date the strains were received and their condition on arrival. The CGC should be acknowledged in any publication that results from the use of strains acquired from the CGC.

3. Preparation of growth media

3.1 Preparation of bacterial food source

Although *C. elegans* can be maintained axenically (1), it is difficult, and the animals grow very slowly. *C. elegans* is usually grown monoxenically in the laboratory using *E. coli* strain OP50 as a food source (2). *E. coli* OP50 is a uracil auxotroph whose growth is limited on NGM plates. A limited bacterial lawn is desirable because it allows for easier observation and better mating of the worms. A starter culture of *E. coli* OP50 can be obtained from the CGC or can be recovered from worm plates (see *Protocol 1*).

Protocol 1.   Preparation of bacterial food source

Equipment and reagents

- Starter culture of *E. coli* OP50 from CGC or a culture recovered from worm plates.
- LB agar: 10 g Bacto-tryptone, 5 g Bacto-yeast, 5 g NaCl, 15 g agar, H_2O to 1 litre. pH 7.5. Sterilize by autoclaving.
- L broth: 10 g Bacto-tryptone, 5 g Bacto-yeast, 5 g NaCl, H_2O to 1 litre. Adjust to pH 7.0 with 1 M NaOH. Sterilize by autoclaving.
- Sterile Petri plates
- 250 ml screw-cap bottles

Method

1. Streak the starter culture or bacteria from a worm plate on to sterile LB agar plates to generate single bacterial colonies (3).

2. Dispense 100 ml aliquots of a rich culture medium, e.g. L broth, into 250 ml screw-cap bottles and autoclave. (The bottles can be stored at room temperature for several months (3).)

3. Use a single colony from the streak plate to aseptically inoculate a bottle of L broth. Incubate at 37 °C overnight.

Protocol 1. *Continued*

4. Store the *E. coli* OP50 streak plates and liquid cultures at 4°C. (They remain usable for several months.)
5. Use the liquid cultures to seed NGM plates (see *Protocol 2*).

3.2 Preparation of NGM Petri plates

C. elegans is maintained in the laboratory on Nematode Growth Medium (NGM) agar, which has been aseptically poured into Petri plates (2). Several sizes of Petri plates are available and can be purchased from companies such as Nunc or Falcon. Smaller plates (35 mm diameter) are useful for matings or when using expensive drugs. Medium-size plates (60 mm diameter) are useful for general strain maintenance, and larger plates (100 mm diameter) are useful for growing larger quantities of worms, such as for certain mutant screens. The NGM agar medium can be poured into Petri plates easily and aseptically using a peristaltic pump. Such a pump, although not essential, is particularly useful when large numbers of plates are required. The CGC uses a Wheaton Unispense liquid dispenser (Wheaton Science Products). This pump can be adjusted so that a constant amount of NGM agar is dispensed into each Petri plate (25 ml for 100 mm plates, 12 ml for deep 60 mm plates, 4 ml for 35 mm plates). A constant amount of agar in the plates reduces the need for refocusing the microscope when you switch from one plate to another. When desired, drugs (e.g. levamisole, streptomycin, or nystatin) can be added to the NGM solution just prior to its being poured (4).

Protocol 2. Preparation of NGM plates

Equipment and reagents

- NaCl
- Agar (e.g. Sigma A1296)
- Bacto-Peptone
- 1 M $CaCl_2$: 55.5 g $CaCl_2$ in 1 litre H_2O. Sterilize by autoclaving.
- 5 mg/ml cholesterol in ethanol (do not: 120.4 g MgSO4 (anhydrous) in 1 litre H_2O. Sterilize by autoclaving.

- 1 M $MgSO_4$: 120.4 g $MgSl_4$ (anhydrous) in 1 litre H_2O. Sterilize by autoclaving.
- 1 M KPO_4 buffer: 108.3 g KH_2PO_4, 35.6 g K_2HPO_4, H_2O to 1 litre, pH 6.0. Sterilize by autoclaving.
- Petri plates
- Peristaltic pump
- 55°C water bath

Method

1. Mix 3 g NaCl, 17 g agar, and 2.5 g peptone in a 2 litre Erlenmeyer flask. Add 975 ml H_2O. Cover the mouth of the flask with aluminium foil. Autoclave for 50 min.

2. Cool the flask in a 55°C water bath for 15 min.

3. Add 1 ml 1 M $CaCl_2$, 1 ml 5 mg/ml cholesterol in ethanol, 1 ml 1 M $MgSO_4$, and 25 ml 1 M KPO_4 buffer. Swirl to mix well.

4. Under sterile conditions, dispense the NGM solution into Petri plates using a peristaltic pump. Fill the plates two-thirds full of agar.

5. Leave the plates at room temperature for 2–3 days before use. This allows contaminants to be detected and excess moisture to evaporate. Store plates in an airtight container at room temperature; they can then be used for several weeks.

It may be desirable to use 96- or 24-well microtitre plates when using expensive drugs or screening a large number of individual animals. For 24-well plates, use 1.5 ml NGM agar and seed with an *E. coli* OP50 lawn (5). If 96-well plates are used, they can be filled with 50 μl of a 1% (w/w) suspension of *E. coli* OP50 in S medium (see *Protocol 3*) (6). However, 96-well plates can also be filled with 50 μl of NGM liquid with *E. coli* HB101 in each well. To do this, packed *E. coli* HB101 cells are suspended at four times their volume in NGM liquid, assuming 1 g of cells equals 1 ml. One volume of the *E. coli* suspension is then added to 24 volumes of NGM liquid (Ralph Clover, personal communication). *E. coli* HB101 is available from the CGC. Care should be taken to prevent worms from crawling between wells; only one strain of *C. elegans* should be used per microtitre plate so that there is no possibility of strains mixing.

3.3 Seeding NGM plates

Using a sterile technique, apply approximately 0.05 ml of an *E. coli* OP50 liquid culture to small- or medium-sized NGM plates or 0.1 ml to large NGM plates using a pipette. If desired, the drop can be spread using the pipette tip or a glass rod, but do not damage the agar surface or the worms will tend to burrow into the agar. Spreading will create a larger lawn, which can aid in visualizing the worms. Take care not to spread the lawn all the way to the edges of the plate; keep the lawn in the centre. The worms tend to spend most of their time in the bacterial lawn. If the lawn extends to the edges of the plate the worms may crawl up the sides of the plate, dry out, and die. Allow the *E. coli* OP50 lawn to grow overnight at room temperature or at 37 °C for 8 hours (cool the plates to room temperature before adding the worms). Seeded plates stored in an airtight container will remain usable for 2–3 weeks.

4. Culturing *C. elegans* on Petri plates

4.1 Transferring worms grown on NGM plates

C. elegans is transparent and can be visualized using a dissecting stereo-microscope equipped with a transmitted light source. The CGC uses Wild Leitz model M5A or Zeiss model SV6. Standard 10 × eyepieces and objectives ranging from 0.6 × to 5 × (total magnification of 6 × to 50 ×) are widely used.

Several methods are used for transferring *C. elegans* from one Petri plate to another. A sterilized scalpel or spatula can be used to move a chunk of agar from an old plate to a fresh plate. There will usually be hundreds of worms in this chunk of agar. The worms will crawl out of the chunk and spread out on to the bacterial lawn of the new plate. This method works well for transferring worms that have burrowed into the agar or are difficult to pick up individually (such as on a starved plate). This method is fine for transferring homozygous stocks, but it is not advisable if the population is heterozygous or if a stock must be maintained by mating.

Another method that will transfer many worms is to use 0.6–1.2 cm by 5.0–7.5 cm strips of sterilized filter paper. The sterilized filter paper is gently set upon the Petri plate, where it absorbs moisture and picks up worms. The filter paper is then touched to a fresh NGM plate where the worms are deposited. Discard the filter paper after use. This method is also fine for transferring homozygous stocks, but it is not advisable if the population is heterozygous or must be maintained by mating.

A third method is to pick up single animals with a worm picker. A worm picker can be made by mounting a 2.5-cm piece of 32-gauge platinum wire into either the tip of a Pasteur pipette or into a bacteriological loop holder. Platinum wire heats and cools quickly and can be flamed often (between transfers) to avoid contaminating the worm stocks. The end of the wire, used for picking up worms, can be flattened slightly with a hammer and then filed with an emery cloth to remove sharp edges; sharp points can poke holes in the worms and kill them or make holes in the agar. The tip of the wire can be fashioned to your liking. Some people prefer a flattened end, while others prefer a slight bend that forms a hook. It takes a bit of experience with a worm picker to avoid poking holes in the agar. Worms crawl into the holes, making it difficult to see or pick them up.

To pick a worm identified under the dissecting microscope, slowly lower the tip of the wire and gently swipe the tip at the side of the worm and lift it up. Another method is to collect a drop of *E. coli* OP50 on the end of the picker before gently touching it to the top of the chosen worm. The worm will stick to the bacteria, provided the bacteria are taken from a seeded plate that is at least two days old. Several animals at a time can be picked by this method, although worms left too long on the pick will desiccate and die. To put a picked worm on a fresh plate, slowly lower the tip of the worm picker, gently touch the surface of the agar, and hold it there to allow the worm to crawl off the picker.

4.2 Transferring frequency

How often you need to transfer your worms depends on their genotype, the temperature at which you grow them, and what you plan to do with them. Heterozygotes and mating stocks should be transferred every one to three generations; it is easier to transfer them before the plates have become starved. If you want to transfer individual animals from a starved plate you may find it easier to first transfer an agar chunk and then to pick individual animals after they have crawled out of the agar chunk and into the bacterial lawn. If the animals are in the dauer stage (see Chapter 1) you may need to wait a day before picking; the worms will develop out of the dauer stage, and the phenotypes can be scored. When transferring heterozygous stocks that are not well balanced it is best to transfer just one worm to each new plate. This allows you to score the progeny of the transferred worm and confirm that you did indeed transfer a heterozygote. Most stocks are maintained as purely self-fertilizing hermaphrodites, but to maintain mating stocks (stocks with both males and hermaphrodites) transfer six to eight adult males and three to four young adult hermaphrodites per plate. You may allow homozygous stocks to starve for several weeks between transfers. Stocks grown at 25 °C will starve several days sooner than identical stocks grown at 16 °C. It is possible to keep stock plates at 11 °C or 16 °C for several months between transfers. To prevent a plate from drying out, wrap it in a strip of Parafilm. Alternatively, you may want to transfer your stocks every 1–2 days so that you have a source of animals at every stage of development.

4.3 General stock maintenance tips

C. elegans stocks can best be maintained between 16 °C and 25 °C, most typically at 20 °C. *C. elegans* grows 2.1 times faster at 25 °C than at 16 °C, and 1.3 times faster at 20 °C than at 16 °C (7). This variation in growth periods can be useful when planning experiments.

Take note of specific growth requirements (e.g. temperature sensitivity or dauer defective) and maintain your stocks accordingly. Such information is included on the Strain Information Sheet sent with each strain shipped from the CGC.

Stock plates do not need to be stored in a covered container. They will eventually dry out, but this won't be a problem if you transfer animals before the plates reach that condition. If you do store your plates in a covered container be sure to watch for signs of high humidity. Although rare, it is possible for worms to crawl (or be carried in a drop of water) from one plate to another in such conditions, resulting in the mixing of strains.

When pouring or seeding plates, or when transferring worms, it is best to keep the amount of time the cover is removed to a minimum. This cuts down on contamination. If you have mould on a plate, carefully wrap the plate with a strip of Parafilm and then dispose of it. Do not remove the cover from a contaminated plate or you risk contaminating other plates. If you must transfer from a contaminated plate, see Section 6.

Label the bottom of the Petri plate with the strain name and date. Do not just label the cover, as it may accidentally be separated from the bottom. Discarded plates should be placed in a plastic bag and autoclaved.

The exponential growth of a population on an agar plate means judging the time of harvesting of the worms just prior to the bacterial food supply being

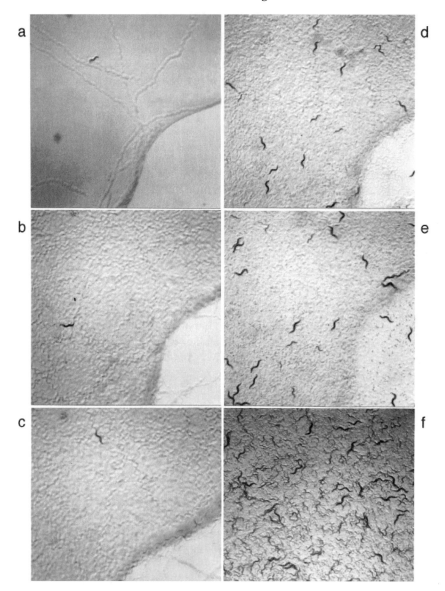

Figure 1. Exponential growth of a plate culture of *C. elegans*; a-f. Views on successive days of the same region of a 5-cm NGM plate seeded with OP50 bacteria. An L4 larva was placed on the plate a few minutes before the first photograph was taken, panel a (day 1), and the plate was then incubated at 20 °C. The small progeny larvae in panel c (day 3), are not clear in the photograph, at this magnification, but become obvious by the following day, panel d (day 4). The bacterial lawn does not extend to the bottom right hand corner of the panels, and eggs are apparent away from the bacterial lawn in panel e. By day 6 (panel f) the bacterial lawn has been cleared by the worm population. All photographs were taken at 6.4× magnification. (Micrographs courtesy of Ian Hope.)

used up, and this takes a little experience. Once the population is large enough the worms sweep across the plate clearing all the bacteria in a few hours (see *Figure 1*).

5. Growth of *C. elegans* in liquid medium

Large quantities of *C. elegans* can be grown in liquid medium. Liquid cultures of *C. elegans* are usually grown in S medium using concentrated *E. coli* OP50 as a food source (8). It is often best to grow just one generation of worms in liquid before the worms are harvested. When growing worms for more than one generation, overcrowding can often lead to dauer formation despite the presence of food.

Overnight cultures of *E. coli* OP50 grown in LB (3) or other rich broth medium should be used to make a concentrated pellet of bacteria. The concentrated pellet can be stored at 4°C for several weeks or in a –70°C freezer indefinitely. It is a good idea to have a large amount of concentrated *E. coli* OP50 available; the amount needed will depend on the starting inoculum of worms and the length of time the worms are grown. As a point of reference, to make a 250 ml batch of a liquid culture of *C. elegans*, add packed *E. coli* OP50 made from a 2 litre culture to 250 ml of S medium and inoculate with worms recovered from four large (100 mm) Petri plates and grow for 4–5 days. The worm culture should be monitored, and additional *E. coli* OP50 added as needed (see *Protocol 3*).

Protocol 3. Preparation of a liquid culture of *C. elegans*

Equipment and reagents

- S basal: 5.85 g NaCl, 1 g K_2HPO_4, 6 g KH_2PO_4, 1 ml cholesterol (5 mg/ml in ethanol), H_2O to 1 litre. Sterilize by autoclaving.
- 1 M potassium citrate pH 6.0: 20 g citric acid monohydrate, 293.5 g tri-potassium citrate monohydrate, H_2O to 1 litre. Sterilize by autoclaving.
- Trace metals solution: 1.86 g disodium EDTA, 0.69 g $FeSO_4 \cdot 7H_2O$, 0.2 g $MnCl_2 \cdot 4H_2O$, 0.29 g $ZnSO_4 \cdot 7H_2O$, 0.025 g $CuSO_4 \cdot 5H_2O$, H_2O to 1 litre. Sterilize by autoclaving and store in the dark.
- 1 M $CaCl_2$. (See *Protocol 2*).

- 1 M $MgSO_4$. (See *Protocol 2*).
- S medium: 1 litre S basal, 10 ml 1 M potassium citrate pH 6, 10 ml trace metals solution, 3 ml 1 M $CaCl_2$, 3 ml 1 M $MgSO_4$. Add components using sterile technique; do not autoclave.
- Four large plates of *C. elegans*, just cleared of bacteria
- Concentrated *E. coli* OP50
- 1–2 litre sterile flask (see step 4)
- 50 ml sterile conical centrifuge tube and centrifuge
- Flask shaker

Method

1. Add 250 ml of the S medium to a sterilized 1–2 litre flask.

2. Inoculate the S medium with a concentrated *E. coli* OP50 pellet [made from 2–3 litres of an overnight culture of E. coli OP50 grown in LB (3) (spin for 10 min at 3000 g at 4°C)].

Protocol 3. *Continued*

3. Wash each of the four large plates of *C. elegans* (just cleared of bacteria) with 5 ml S medium and add to the 1–2 litre flask.

4. Put the flask on a shaker at 20°C. Use fairly vigorous shaking or fluted flasks so that the culture is well oxygenated.

5. Monitor the cultures by checking a drop of the culture under the microscope. Add more concentrated *E. coli* OP50 suspended in S medium if the food supply is depleted (i.e. the solution is no longer visibly cloudy). Harvest the culture when there are many adult animals in each drop—this is usually on the 4th or 5th day.

6. Put the flask on ice for 15 min to allow the worms to settle.

7. Aspirate most of the liquid from the flask.

8. Transfer at room temperature the remaining liquid to a 50 ml sterile conical centrifuge tube and spin for at least 2 min at 1150 *g* to pellet the worms. Young larvae may take longer than 2 min to pellet.

9. Aspirate the remaining liquid.

The worms harvested from growth in liquid culture are usually longer and thinner than those grown on Petri plates, and tend to retain their eggs. The number of adults, larvae, and eggs obtained depends on the strain of worms used, the amount of bacteria provided, and the length of time the culture is grown. Freshly cleared cultures will yield worms equal to half the weight of the bacteria used (8).

6. Cleaning contaminated *C. elegans* stocks

Occasionally, *C. elegans* stocks may become contaminated with other bacteria, yeast, or mould. It is easy to rid your worm stocks of contaminants. Most contaminants will not hurt the worms. (In fact, the worms like to climb some fungal hyphae and sway back and forth!) But it is much easier to score phenotypes and perform transfers when your stocks are clean, and some contaminants cause the worms to burrow into the agar. Mould can be removed by serially transferring and allowing the worms to crawl away from the contaminant. Bacterial contaminants and yeast are easily removed by treating with a hypochlorite solution, which will kill the contaminant and all worms not protected by the egg shell. This can be done using an entire contaminated plate, or it can be done quickly using a single hermaphrodite (technique modified from ref. 9).

Protocol 4. Removing mould contaminants from *C. elegans* stock plates

Method

1. Sterilize a scalpel or spatula in a flame and remove a chunk of agar from the contaminated plate. Remove the cover of the contaminated plate only as long as necessary.

2. Place the chunk of agar at the edge of a seeded clean plate. Allow the worms to crawl out of the chunk and across the *E. coli* OP50 lawn to the opposite side of the plate.

3. Once the worms have reached the other side of the plate, pick up individual animals with a worm picker (or take another chunk using a flamed scalpel or spatula) and place it on the edge of another clean plate.

4. Repeat step 3.

Protocol 5. Egg preparation: removing bacterial or yeast contaminants from *C. elegans* stock plates

Equipment and reagents

- 5 M NaOH
- Household bleach (5% solution of sodium hypochlorite)
- Sterile 15 ml capped conical centrifuge tube
- Shaker or vortex mixer
- Table-top centrifuge
- Sterile Pasteur pipettes
- Two clean NGM plates, each seeded with an *E. coli* OP50 lawn (see *Protocol 2*)

Method

1. Use contaminated *C. elegans* stock plates that have many gravid hermaphrodites.[a] Wash the plates with sterile H_2O. Pipette the H_2O across the plate several times to the loosen worms and eggs that are stuck in the bacteria.

2. Collect the liquid and worms in a sterile 15 ml conical centrifuge tube with a cap. Add H_2O to a total volume of 3.5 ml.

3. Mix 0.5 ml 5 M NaOH with 1 ml bleach. Make this solution fresh just before use! Add to the centrifuge tube containing the worms.

4. Shake well or vortex the tube for a few seconds. Repeat shaking/vortexing every 2 minutes for a total of 10 minutes.

5. Centrifuge in a table-top centrifuge for 30 sec at 1300 *g* to pellet the released eggs.

6. Aspirate to 0.1 ml.

Protocol 5. *Continued*

7. Add sterile H$_2$O to 5 ml. Shake well or vortex for a few seconds.
8. Repeat steps 5 and 6.
9. Use a sterile Pasteur pipette to transfer the eggs in the remaining 0.1 ml of liquid to the edge of a clean NGM plate seeded with an *E. coli* OP50 lawn.
10. The next day the eggs will have hatched and the larvae will have crawled into the *E. coli* OP50 lawn. Transfer the hatched larvae to a clean NGM plate seeded with a lawn.

[a] It is embryos inside eggshells that will survive this procedure.

Protocol 6. Egg preparation in a drop: small-scale method for removing bacterial or yeast contaminants

Equipment and reagents

- 1 M NaOH
- Household bleach (5 % solution of sodium hypochlorite)
- Two clean NGM plates, each seeded with an *E. coli* OP50 lawn (see *Protocol 2*)

Method

1. Make a 1:1 mixture of 1 M NaOH:bleach. Put a drop of this solution on the edge of a clean NGM plate seeded with an *E. coli* OP50 lawn.
2. Put several gravid hermaphrodites in the drop. The solution will kill the contaminants and hermaphrodites but will soak into the plate before the embryos hatch.
3. The next day the larvae will have crawled into the *E. coli* OP50 lawn. Transfer them to a clean NGM plate seeded with an *E. coli* OP50 lawn.

Be very careful that your stocks do not become contaminated with mites. Mites can crawl from plate to plate, transferring worm eggs and larvae on their feet. This will result in the cross-contamination of your stocks. If you do develop a mite problem, locate the source of the mites and remove it. Parafilm all contaminated plates and autoclave. Vapour from a beaker of moth crystals (paradichlorobenzene) placed in the incubator will kill the mites without affecting the worms.

7. Obtaining synchronous cultures and staged animals

Synchronously growing cultures of *C. elegans* can be obtained by starting with the eggs recovered from the egg preparation procedure (see *Protocol 5*). The eggs are allowed to hatch overnight in the absence of food, and the resulting culture will consist of starved worms arrested in the L1 stage of development. Introducing food into the culture allows the worms to resume development, all from the same starting age.

Protocol 7. Obtaining synchronous cultures of *C. elegans*

Equipment and reagents

- M9 buffer: 3 g KH_2PO_4, 6 g Na_2HPO_4, 5 g NaCl, 1 ml 1 M $MgSO_4$, H_2O to 1 litre. Sterilize by autoclaving.
- S basal (see *Protocol 3*)
- Concentrated *E. coli* OP50 (see *Protocol 3*, Step 2)

- Axenized *C. elegans* eggs from four to eight 100 mm plates (see *Protocol 5*)
- Two sterile 1–2 litre flasks
- 50 ml sterile conical centrifuge tube and centrifuge.

Method

1. Aseptically transfer the axenized eggs to 250 ml of the M9 buffer in a 1–2 litre flask and allow to incubate overnight at 20°C, using fairly vigorous shaking, to obtained starved L1 animals.
2. Put the flask on ice for 15 min to allow the worms to settle.
3. Aspirate most of the liquid from the flask (to remove any dauer pheromone accumulated during starvation).
4. Transfer the remaining liquid to a 50 ml sterile conical centrifuge tube and centrifuge for at least 2 min at 1150 *g* to pellet the worms.
5. Aspirate the remaining liquid.
6. Transfer the worms to 250 ml of the S basal inoculated with concentrated *E. coli* OP50 in a 1–2 litre flask. Place flask on a shaker at 20°C. Use fairly vigorous shaking so that the culture is well oxygenated.
7. Monitor by checking a drop of the culture under the microscope. Add more food as necessary.[a]

[a] At 20°C mid-L1 larvae can be harvested after approximately 8 h, mid-L2 larvae at approximately 18 h, mid-L3 larvae at approximately 25 h, and mid-L4 larvae at approximately 37 h (10).

If only a small number of synchronized animals are needed, the eggs can be added to a thin layer of M9 buffer in a 60 mm Petri plate and allowed to hatch overnight. The starved L1 animals can then be placed on seeded NGM Petri plates, and synchronous growth will begin with the reintroduction of food. Animals should be monitored using the microscope and harvested at the desired stages.

A population of dauer animals can be obtained by adding dauer-inducing pheromone to the liquid culture (11). Dauers can also be obtained by allowing a liquid culture to grow for several days after the culture is cleared of bacteria. The dauers can be separated from other stages by treating them with 1% sodium dodecyl sulfate (SDS) for 30 min in a rotating tube and then washing once with H_2O and once with M9 buffer.

8. Freezing and recovery of *C. elegans* stocks

Caenorhabditis elegans can be frozen and stored indefinitely in liquid nitrogen (–196°C) (2). The keys to a successful freeze are using animals at the correct stage of development, the addition of glycerol to the freezing medium, and a gradual cooling to –80°C. Freshly starved young larvae (L1–L2 stage) survive freezing best. Well-fed animals, adults, eggs, and dauers do not survive well. It is best to use several plates of worms that have just exhausted the *E. coli* OP50 lawn and that contain many L1–L2 animals. A 15% final volume of glycerol in the freezing solution is used. A 1°C decrease in temperature per minute is desirable during freezing. This can be achieved by placing the worms (in freezer vials) in a styrofoam container at –80°C. The styrofoam container can be either a commercial shipping box (with walls at least 75 mm thick) or a small styrofoam box with slots for holding vials. After 12 or more hours at –80°C, the freezer vials should be transferred to their permanent freezer location for long-term storage.

The CGC uses two solutions for freezing *C. elegans*: a Liquid Freezing Solution (2) and a Soft Agar Freezing Solution (Leon Avery, personal communication). For the long-term storage of stocks in liquid nitrogen, the Liquid Freezing Solution is recommended. When this solution is used the worms settle to the bottom of the freezer vial, and no viable animals can be easily retrieved without thawing the entire contents of the vial. The Soft Agar Freezing Solution is useful for freezing working stocks of *C. elegans*. The addition of the agar helps to keep the worms suspended throughout the solution. A small scoop of the frozen contents can be taken, and the remainder can be left in the vial and returned to the freezer for later use. Vials frozen using the Soft Agar Freezing Solution should be stored at –80°C. If kept in liquid nitrogen, the contents become too hard, and the vial will need to be warmed before it is possible to remove a scoop of the contents. The warming period reduces the number of times live worms can be recovered from the vial. When stored at –80°C there is no need to warm the vial; the contents are soft enough to be used right away. We recover worms three to four times from each vial of soft agar stock.

Worms can be frozen in 1.8 ml capacity cryotubes. The CGC uses Nunc cryotube vials (cat. no. 65234) with internal threads, and the CGC freezes six vials of each strain it receives. Two vials are frozen using the Soft Agar Freezing Solution and these are stored at –80°C for use as working stocks. The other four vials are frozen using the Liquid Freezing Solution: one is thawed as a tester, and the other three are put in at least two different liquid nitrogen tanks. To fill strain requests, the working stocks kept at –80°C are used. When the last vial of a stock stored at –80°C is emptied, the worms are once again frozen using the Soft Agar Freezing Solution in order to replace these vials. In theory, the stocks kept in liquid nitrogen will never need to be used since the soft agar stocks are continually replaced as they are depleted.

The recovery of *C. elegans* from stocks stored in liquid nitrogen is in the range from 35% to 45% of the total number of animals frozen. This number decreases only slightly after many years of storage in liquid nitrogen. The recovery of stocks stored at $-80\,^{\circ}C$ for many years (>10) is not as high as those kept in liquid nitrogen, but worms can be safely stored this way for many years (CGC, unpublished data). Of course, a power failure can result in the loss of all stocks kept at $-80\,^{\circ}C$, so it is very wise to keep at least one copy of all stocks in liquid nitrogen. Some mutants strains (especially certain *Dpy* mutants) do not survive freezing as well as wild-type animals.

Protocol 8. Freezing *C. elegans* using the Liquid Freezing Solution

Equipment and reagents
- NGM plates of freshly starved L1–L2 animals
- Sterile test tubes
- S buffer: 129 ml 0.05 M K_2HPO_4, 871 ml 0.05 M KH_2PO_4, 5.85 g NaCl. pH 6.0. Sterilize by autoclaving.
- S buffer + 30% glycerol (v/v). Sterilize by autoclaving.
- 1.8 ml cryotube vials
- Styrofoam box
- $-80\,^{\circ}C$ freezer

Method
1. Use one large, two to three medium, or five to six small NGM plates that contain lots of freshly starved L1–L2 animals. Wash the plates with 0.6 ml S buffer for each vial you will freeze. Collect the liquid in a sterile test tube.
2. Add an equal volume of S buffer + 30% glycerol. Mix well.
3. Aliquot 1.0 ml of the mixture into 1.8 ml cryovials labelled with the strain name and date.
4. Pack the cryovials in a small styrofoam box with slots for holding microtubes or use a commercial styrofoam shipping box.
5. Place the box in a $-80\,^{\circ}C$ freezer overnight (or for at least 12 h).
6. The next day transfer the vials to their permanent freezer locations. Thaw one vial as a tester to check how well the worms survived the freezing (see *Protocol 10*).

Protocol 9. Freezing *C. elegans* using the Soft Agar Freezing Solution

Equipment and reagents
- S buffer (see *Protocol 8*)
- Soft Agar Freezing Solution: 0.58 g NaCl, 0.68 g KH_2PO_4, 30 ml glycerol, 0.87 g K_2HPO_4, 0.4 g agar, H_2O to 100 ml. pH 6.0. Sterilize by autoclaving.
- NGM plates
- Sterile test tubes
- 1.8 ml cryotube vials
- Styrofoam box
- $50\,^{\circ}C$ water bath

Protocol 9. *Continued*

Method

1. Melt the Soft Agar Freezing Solution in an autoclave or microwave and place in a 50°C water bath for at least 15 min.

2. Use one large, two to three medium, or five to six small NGM plates that contain lots of freshly starved L1–L2 animals. Wash the plates with 0.6 ml S buffer for each vial you will freeze. Collect the liquid in a covered sterile test tube and place in ice for 15 min.

3. Add an equal volume of the Soft Agar Freezing Solution to the test tube. Mix well.

4. Aliquot 1 ml of this mixture into 1.8 ml cryovials labelled with the strain name and date.

5. Pack the cryovials in a small styrofoam box with slots for holding microtubes, or use a commercial styrofoam shipping box.

6. Place the box in a –80°C freezer overnight (or for at least 12 h).

7. The next day transfer the vials to their permanent freezer locations. Take a scoop of the frozen mixture from one vial as a tester to check how well the worms survived the freezing (see *Protocol 11*).

Protocol 10. Thawing *C. elegans* frozen using the Liquid Freezing Solution

Method

1. Remove a vial from the freezer and allow it to thaw at room temperature until all the ice has turned to liquid.

2. Pour the contents on to one large NGM plate with an *E. coli* OP50 lawn. You should see worms wiggling after just a few minutes.

3. After 2–3 days, transfer 10–15 animals individually to separate plates. Allow the animals to reproduce for one generation and score the progeny for correct phenotypes.

Protocol 11. Thawing *C. elegans* frozen using the Soft Agar Freezing Solution

Method

1. Remove a vial from the –80°C freezer and transfer to a small styrofoam box with slots for microtubes.

2. Work quickly so that the solution in the vial does not thaw. Flame a small scoop or spatula and use it to remove 0.25–0.33 ml of the frozen solution. Place the solution on a NGM plate with an *E. coli* OP50 lawn. You should see worms moving after just a few minutes.

3. Return the vial to the –80°C freezer as quickly as possible.

4. After 2–3 days, transfer 10–15 animals individually to separate plates. Allow the animals to reproduce for one generation and score the progeny for correct phenotypes.

Acknowledgements

The author wishes to thank Robert Herman, Ann Rougvie, Todd Starich, and Ian Hope for their comments.

References

1. Vanfleteren, J. R. (1980). In *Nematodes as Biological Models*, Vol. 2 (ed. B. M. Zuckerman), p. 47. Academic Press, New York.
2. Brenner, S. (1974). *Genetics*, **77**, 71.
3. Maniatis, T., Fritsch, E. F., and Sambrook, J. (1982). *Molecular Cloning: A Laboratory Manual*, p. 68. Cold Spring Harbor Laboratory Press, New York.
4. Avery, L. (1993). *Genetics*, **133**, 897.
5. Hartman, P. S. and Herman, R. K. (1982). *Genetics*, **102**, 159.
6. Rogalski, T. M. and Riddle, D. L. (1988). *Genetics*, **118**, 61.
7. Hirsh, D., Oppenheim, D., and Klass, M. (1976). *Dev. Biol.*, **49**, 200.
8. Lewis, J. A. and Fleming, J. T. (1995). In *Methods in Cell Biology*, Vol. 48 (ed. H. F. Epstein and D. C. Shakes), p. 3. Academic Press, San Diego, CA.
9. Sulston, J. and Hodgkin, J. (1988). In *The Nematode* Caenorhabditis elegans (ed. W. B. Wood), p. 587. Cold Spring Harbor Laboratory Press, New York.
10. Byerly, L., Cassada, R. C., and Russell, R. L. (1976). *Dev. Biol.*, **51**, 23.
11. Caldicott, I. M., Larsen, P. L., and Riddle, D. L. (1994). In *Cell Biology: A Laboratory Handbook*, p. 389 Academic Press, San Diego, CA.

Transformation

YISHI JIN

1. Introduction

The technique of transformation was introduced to *C. elegans* in the early 1980s, when tRNAs isolated from *sup-7(st5)* amber suppressor mutants were injected into the gonads of *tra-3* animals to suppress the sex transformation phenotype of an amber suppressible mutation, *tra-3(e1107)* (1). Soon after, DNA-mediated germline transformation was evolved both in integrative and non-integrative forms (2, 3).

By injecting DNA containing the *sup-7(st5)* mutant DNA into appropriate *C. elegans* hosts, Fire obtained transformants in which the foreign DNA was maintained in low copy numbers. The foreign DNA was integrated into the genome at a relatively high frequency (2, 4). However, the DNA had to be injected into the nuclei of oocytes, which is technically difficult, and hence this method has not been widely used.

The most popular injection procedure was developed by Stinchcomb, Mello, and their co-workers (3, 5). In this technique, DNA is injected into the cytoplasm of the syncytial, mitotically active, gonad. The injected DNA contains a selectable marker to facilitate the identification of transformants in the progeny. Usually, after following the selectable co-injection marker for two generations, the injected DNA can be stably passed through the germline in the form of extrachromosomal arrays that resemble free chromosomal duplications.

Transformation of *C. elegans* has been used for several purposes: to identify a gene by rescuing a mutant phenotype using a wild-type copy of the DNA; to analyse the expression pattern of a gene of interest using reporter constructs; to interfere with a biological process by overexpression of a wild-type or mutant form of a gene of interest; and to analyse the site of gene action by mosaic analysis. In this chapter I describe the essential elements of the technique, the current understanding of the fate of DNA following transformation, and common uses for this technique. A detailed description of transformation can also be found in ref. 6.

2. The *C. elegans* germline

The gonad includes the somatic gonad and germline, and is the structure with which one must be familiar for transformation. The development of the hermaphrodite gonad is illustrated in *Figure 1*. The mature hermaphrodite gonad consists of two U-shaped tubular arms that have distinct appearances under Nomarski optics as shown in *Figure 2*. The details of germline development can be found in refs 7 and 8. A brief overview is given below.

When a worm hatches from the egg, the gonad primordium is composed of four cells: Z1 and Z4 are precursors for the somatic gonad; Z2 and Z3 are the germline precursors. During early larval (L1) development the somatic gonad precursors undergo cell division. In the late L2 stage, the somatic gonad primordium is formed to enclose the germ cells into anterior and posterior arms, with the centre around the future vulva. The gonad sheath, sperma-theca, and uterus are produced from somatic gonad cells by the L3 and L4 stages. Germline precursors divide throughout all larval and adult stages,

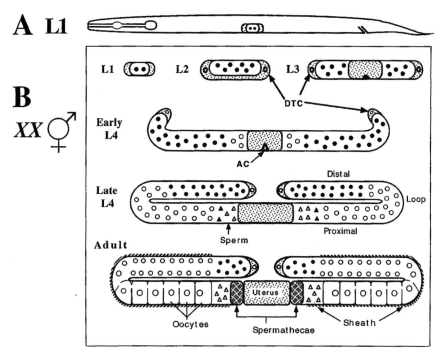

Figure 1. Development of the hermaphrodite gonad. (A) Schematic diagram of a newly hatched, first-stage larva. The gonad primordium is located midventrally. (B) Schematic diagram of the development and anatomy of the gonad at selected larval (L) and adult stages. Somatic gonad cells are shaded. Germline cells are: ● for mitotic nuclei; ○ for meiotic prophase nuclei; ▲ for primary spermatocytes; △ for sperm. DTC, distal tip cells; AC, anchor cell. Anterior is left. (Reproduced, with permission, from ref. 8.)

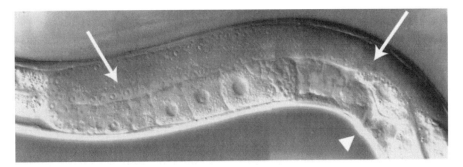

Figure 2. Nomarski photograph of the anterior gonad arm of a young adult hermaphrodite. Dorsal is up. Arrowhead, the vulva. Left arrow, the distal region of the gonad syncytium. Right arrow, the proximal portion. (Modified, with permission, from ref. 8.)

generating more than 1000 nuclei per gonad arm in the mature hermaphrodite. Morphogenesis of the gonad is led by two somatic cells located at the distal end of each gonad arm, called distal tip cells (DTCs). DTCs first migrate along the anteroposterior body axis away from the centre, thus elongating the gonad. They then turn dorsally and migrate towards the centre, thus forming the U-shaped gonad with two arms on the dorsal side of the animal and the central opening on the ventral side to connect to the future vulva (see *Figure 1*).

The distal region of the germline, on the dorsal side and far from the opening of the gonad, is composed of mitotically active germ-cell nuclei that lie within a syncytium. The germ-cell nuclei have a uniform button-like appearance under Nomarski optics (see *Figure 2*), and are arranged peripherally around a central cytoplasmic core that runs along the centre of the gonad tube. Moving towards the opening of the gonad, germ cells exit the mitotic cycle to enter meiosis. Oogenesis starts at the turn (or loop) of each gonad arm, and oocytes mature gradually as they move proximally. Spermatogenesis occurs near the opening of the gonad such that fertilization occurs when oocytes pass through the spermatheca. Fertilized zygotes undergo a few rounds of cell division before being expelled from the vulva.

3. Equipment for transformation

(a) *A standard dissecting microscope*: This is used for the routine manipulation of *C. elegans*, and should be equipped with optics for magnification from 5–10 × to 40–50 × and a transmitted-light illuminator. Several companies including Leica, Zeiss, Nikon, and Olympus, sell stereo microscopes with slight variations. Which kind of microscope to use depends on one's specific needs and financial situation. We have used the Zeiss MS5, which is relatively cheap and satisfactory for most worm manipulations.

For first-time buyers it is helpful to search on the *C. elegans* Web site for valuable tips and discounts (http://elegans.swmed.edu/).

(b) *An inverted compound light microscope*: The most popular ones are the Zeiss Axiovert, Nikon, or Olympus equivalents. It should be equipped with at least two differential interference contrast objectives of 5–10 × (for locating the worm and needle) and 40 × dry (for injection). Because oils used in worm injections present a hazard to the lenses, an empty slot may be arranged in between the two lenses to serve as a resting position underneath the injection needle when one is not injecting. A glide stage is essential for fine alignment. This microscope can also be used for other purposes, such as fluorescence, if one can not afford another microscope.

(c) *A micromanipulator*: The injection needle is held in a needle holder that is attached to the compound microscope via a micromanipulator. The most popular micromanipulators are the Joy-stick series from Narishige (Japan), which can be purchased through the microscope vendors—discounts can be obtained if one buys a microscope and a micromanipulator at the same time. Which kind of micromanipulator to purchase depends on the available finances; fine control along the vertical axis (Z-axis) is the most critical element. A home-made apparatus can be used instead if finances are constrained.

(d) *An injection manipulator connected to a pressurized nitrogen cylinder*: This is not absolutely essential. It can be obtained from Medical System Corp. (Greenvale, New York), Eppendorf (http://www.eppendorf.com/), and Narishige (Japan). Most injection manipulators have three adjustable pressure controls: a pressure less than 5 p.s.i. for resting, a pressure from 10–20 p.s.i. for injection, and a high pressure that directly measures the tank pressure for clearance. In addition, a footpedal attachment can be very convenient for controlling the pressure while injecting. However, experienced injectors often prefer hand-operation systems. On a low budget, this equipment can be replaced with a syringe filled with oil or air that connects to the back of the needle holder through tubing.

(e) *A needle puller*: Reliability and reproducibility are the critical elements in choosing a needle puller. Several brands of needle pullers are available from Sutter (Sutter: http://www.sutter.com/), Narishige (Japan) and other companies. A needle puller can be shared among many laboratories using different sizes of needles.

4. Procedures for transformation

The following procedures are for transformation by injection into the syncytial gonad. They include sample preparation, needle preparation, worm immobilization, injection and recovery, and identification of transformants. Each step contributes to the success of transformation, and can be practised separately. A method using microprobes for transformation in *C. elegans* has been reported (9), but not widely used.

4.1 Preparing DNA samples for transformation

4.1.1 Making DNA

Apparently, *C. elegans* will replicate any DNA introduced into the germline and there are no specific vectors needed. DNA prepared by any standard method can be used in worm transformation. *Protocols 1 and 2* are sample procedures for preparing plasmid and phage DNAs. Some people favour using DNAs prepared by commercial kits, such as Qiagen columns or Wizard mini-prep kits. However, we have found that some commercial DNA mini-prep kits yield DNA that is toxic to worms. Thus, when one chooses a new DNA purification kit, it is wise to check that the DNA preparation does not cause toxicity. RNA contamination in the DNA preparation is not a problem in transformation, so RNase digestion is unnecessary. Plasmid DNA prepared following the procedure of lysis by boiling (10) is toxic to worms. Any phenol remaining after the purification of DNA by phenol–chloroform extraction can also cause toxicity.

Protocol 1. Isolation of plasmid DNA for transformation

Reagents

- 1.5 ml overnight culture of *E. coli* transformed with a plasmid
- Lysis solution: 1% SDS, 0.2 M NaOH
- 7.5 M NH$_4$acetate
- TE buffer: 10 mM Tris–HCl pH 7.5, 1 mM EDTA pH 8.0
- Isopropanol
- 70% ethanol

Method

1. Centrifuge the overnight culture for 30 sec at 12 000–20 000 g and resuspend the *E. coli* cell pellet in 200 μl TE buffer by vigorous vortexing.

2. Add 400 μl of the lysis solution, then invert the tube gently a few times until the mixture goes clear, and leave at room temperature for 3–5 min.

3. Add 300 μl of 7.5 M NH$_4$acetate, invert the tube a few times until there is no separation of layers and white flakes are formed uniformly, and leave on ice for 5–10 min.

4. Centrifuge the tube in a microcentrifuge at 20 000 g (14 000 r.p.m.) for 8 min, and transfer the supernatant to a new tube. Avoid touching the lower part of the liquid.

5. Add 500 μl of isopropanol, mix by inversion and leave at room temperature for 5 min.

6. Centrifuge the tube in a microcentrifuge at 20 000 g (14 000 r.p.m.) for 10 min, and discard the supernatant.

Protocol 1. *Continued*

7. Wash the DNA pellet with 70% ethanol, dry the pellet under vacuum or in the open air. (At this stage you may or may not see a white pellet at the bottom of the tube. It is better not to have any visible pellet.)

8. Dissolve the DNA in 50 μl of TE buffer.

This procedure works well with any bacterial culture. DH5α appears to give the cleanest DNA prep.

If this procedure is used for preparing cosmid DNAs, start with 4 ml of a culture, scale up proportionally, and leave the tubes on ice for longer at step 3.

Protocol 2. Preparation of phage DNA for transformation

Equipment and reagents

- Phage stock
- LB medium: 10 g Bacto-tryptone, 5 g Bacto-yeast extract, 10 g NaCl per litre. Sterilize by autoclaving.
- 1 M MgSO₄ stock solution
- 10% maltose stock solution
- Chloroform
- 10 mg/ml DNase I. Make up fresh.
- 10 mg/ml RNase A
- 40% PEG (polyethylene glycol 8000), 2.5 M NaCl
- 0.5 ml SM: 0.1 M NaCl, 10 mM MgSO₄, 50 mM Tris pH 7.5, 0.01% gelatin
- 2 M Tris pH 8
- 200 mM EDTA pH 8

- 10% SDS
- Phenol:chloroform (1:1 (v/v))
- 3 M sodium acetate pH 5.2
- Isopropanol
- TE buffer: 10 mM Tris pH 7.5, 1 mM EDTA pH 8.0
- 70% ethanol
- Sorvall or Beckman centrifuge, JA-20 rotor for Beckmen, SS-34 rotor for Sorvall
- 30 ml Corex centrifuge tubes (or equivalent)
- 1.5 ml Eppendorf tubes
- Yellow micropipette tips

Method

1. Grow an overnight culture of an appropriate host bacteria in LB medium with 10 mM MgSO₄ and 0.1% maltose.

2. Dilute fresh cells 1:100 in 20 ml LB with 10 mM MgSO₄ and 0.1% maltose in a 250 ml flask.

3. Inoculate with about 10^7 p.f.u. of phage stock.

4. Shake at 37°C until just lysed: the cloudy bacteria culture becomes clear and lysed bacterial debris is seen as stringy material. This should take 4–6 h; if the culture lyses too quickly there will not be enough phage.

5. Add 0.5 ml of chloroform and return to the shaker for 15 min. This will lyse any remaining bacteria. Pour the culture into a 30 ml Corex tube (or equivalent) and centrifuge at 4000 g for 10 min (6000 r.p.m. in a Sorvall or Beckman centrifuge).

6. Carefully decant the supernatant into a new centrifuge tube, avoiding the chloroform droplet.

7. Add DNase I and RNase A to 1 mg/ml, and incubate at 37°C for 30 min.

8. Add 5 ml of 40% PEG (polyethylene glycol 8000), 2.5 M NaCl to give a final concentration of 8% PEG and 0.5 M NaCl, and incubate on ice 30 min. This precipitates the phage particles.

9. Centrifuge at 4000 *g* for 10 min (6000 r.p.m. in a Sorvall or Beckman centrifuge) to pellet the phage particles.

10. Drain the phage pellet well. The pellet may be smeared up the side of the tube.

11. Completely resuspend the pellet in 0.5 ml SM by vortexing.

12. Add 50 μl 2 M Tris (pH 8) and 50 μl 200 mM EDTA (pH 8), and mix.

13. Add 50 μl 10% SDS, and heat to 65°C for 5–10 min to break open the phage.

14. Centrifuge briefly (seconds) to make sure that all the liquid is at the bottom of the tube, and transfer the suspension to a 1.5 ml Eppendorf tube.

15. Extract twice with phenol:chloroform. **Do not vortex**. Mix by inversion until a milky emulsion is seen. Centrifuge in a microcentrifuge at 6000 g (7000–8000 r.p.m.) for 2 min. Remove the upper aqueous phase using cut-off yellow micropipette tips. (Lambda DNA is 40–50 kb long and is concatamerized, therefore it must be treated gently to avoid shearing.)

16. Add 50 μl of 3 M Na acetate (pH 5.2) and 600 μl of isopropanol to precipitate the DNA. You should see a stringy precipitate.

17. Centrifuge in a microcentrifuge for 5 min at 15 000 g (12 000 r.p.m.) and resuspend in 500 μl TE buffer. If the suspension is cloudy you have RNA present; add 1 μl RNase A (10 mg/ml) and digest for 15 min at 37°C.

18. Re-precipitate by adding 150 μl of 3 M Na acetate (pH 5.2) and 400 μl of isopropanol. Leave at room temperature for 10 min then centrifuge in a microcentrifuge for 5 min at 15 000 g (12 000 r.p.m.) and wash the pellet with 70% ethanol. Air-dry the pellet, and resuspend in TE buffer. The resuspension takes a long time because the lambda DNA is concatamerized; either resuspend overnight at 37°C or at 50°C for a few hours.

Note: Lambda DNA can be prepared similarly from phage lysis collected from confluent plates.

One obstacle in cloning genes available only on YAC clones has been the requirement of special electrophoresis apparatus such as pulse-field or CHEF=contour-clamped homogeneous field electrophoresis gel electrophoresis. This may be remedied by a simple procedure (see *Protocol 3*) developed by A. Davies and J. Shaw (personal communication).

Protocol 3. Preparing YAC (+ yeast) DNA for transformation[a]

Equipment and reagents

- Yeast strain carrying the YAC of interest
- Selective medium (e.g. CM URA⁻ TRP⁻)
- YPD medium: 10 g yeast extract, 20 g peptone per 900 ml H_2O (autoclaved) plus 100 ml 20% dextrose (filter sterilized)
- 1 M sorbitol
- 1 M sorbitol, 0.1 M EDTA (pH 8.0), 0.1% β-mercaptoethanol
- 2.5 mg/ml zymolase or 10 mg/ml yeast lytic enzyme
- TE buffer: 10 mM Tris–HCl pH 8.0, 1 mM EDTA (pH 8.0)
- 0.5 M EDTA pH 8.0
- 2 M Tris base (pH not adjusted)
- 10% SDS
- DEPC (diethyl pyrocarbonate)

- 5 M potassium acetate pH 4.8–5.2
- Ethanol
- CsCl
- 10 mg/ml ethidium bromide
- 1 ml syringe with a 20–21 gauge needle
- Isoamyl alcohol
- Sorvall or Beckman centrifuge, JA-20 rotor for Beckmen, SS-34 rotor for Sorvall
- 30 ml centrifuge tubes
- Ultracentrifuge tubes (13 × 51 mm, quick-seal) (Beckman)
- vTi65 rotor (or equivalent)
- UV illuminator and safety equipment

Method

1. Grow a 3 ml culture of the yeast strain carrying the YAC of interest in selective medium (e.g. CM URA⁻ TRP⁻) at 30°C for 1–2 overnights.

2. Add the culture to 50 ml of the YPD medium in a flask and shake overnight at 30°C.

3. Centrifuge the yeast culture at 4000 *g* for 5 min (6000 r.p.m. in a Sorvall or Beckman centrifuge). Wash the yeast pellet with 3.5 ml of 1 M sorbitol, and re-centrifuge the cells.

4. Resuspend the yeast pellet in 3.5 ml of 1 M sorbitol, 0.1 M EDTA (pH 8.0), 0.1% β-mercaptoethanol.

5. Add 0.3 ml zymolase or 0.3 ml yeast lytic enzyme (cheaper but not as effective) and incubate at 37°C for at least 1 h.[b]

6. Centrifuge at 4000 *g* for 3 min (5000–6000 r.p.m. in a benchtop centrifuge). Resuspend in 3.2 ml of TE buffer.

7. Add 0.32 ml of 0.5 M EDTA, 0.16 ml of 2 M Tris base, 0.16 ml of 10% SDS, 10 μl of DEPC. Mix by inversion, then incubate at 65°C for 30 min.

8. Add 0.8 ml of 5 M potassium acetate and chill on ice for about 1 h.

9. Centrifuge at 15000 *g* at 4°C for 10 min.

10. Transfer the supernatant to a 30 ml centrifuge tube, add 12 ml of room-temperature ethanol, mix, and centrifuge at 15000 *g* for 15 min.

11. Resuspend the DNA pellet in 3.0 ml of TE buffer, add 3.87g of CsCl and, make up the volume to 4.9 ml.[c]

12. After the CsCl has dissolved (by swirling), add 0.1 ml of ethidium

bromide. (It may be useful at this point to centrifuge down any precipitate that has formed. This can interfere with the banding process later.)

13. Transfer to ultracentrifuge tubes (13 $\times$ 51 mm, quick-seal tubes), balance the tubes, and heat seal.

14. Centrifuge in a vTi65 rotor or equivalent at 400 000 g (50 000 r.p.m.) at 22°C for 18 h.

15. Under UV illumination recover the DNA band using a 1 ml syringe and a 20–21 gauge needle. (Cutting off the top of the tube or inserting a needle at the top of the tube may help to prevent bubbles disrupting the band.)

16. Dilute the sample with 1.5 volumes of double-distilled (dd) H_2O and extract the ethidium bromide by adding equal volumes of isoamyl alcohol, mixing gently, and removing the alcohol until the colour is gone.

17. Precipitate the DNA with 2.5 volumes of ethanol at room temperature.

18. Centrifuge and resuspend in 50–500 μl TE buffer depending on yield (if your spheroplast step goes well then 300–500 μl is more appropriate).

19. The YAC DNA (+yeast genomic DNA) can be injected at concentrations from 15 μg/ml to 150 μg/ml.

[a] Provided by Andrew Davies and Jocelyn Shaw, University of Minnesota; e-mail: davies@biosci.cbs.umn.edu, or jocelyn@biosci.cbs.umn.edu
[b] It is important to check the cells before continuing. This treatment should generate spheroplasts that will burst in distilled H_2O (check under a microscope). If a high proportion of spheroplasts is not obtained, wait longer or add more enzyme.
[c] The Cs gradient purification, steps 11 to 18, may be avoided by using Qiagen columns.
1. Follow the procedure above to the end of step 10, the point of resuspending the DNA from spheroplasts
2. Resuspend the DNA pellet in 2.5 ml of TE buffer, but do not add the CsCl or ethidium bromide.
3. Add 25 μl RNase A (10 mg/ml) and incubate for 10 min at 37°C.
4. Purify the DNA through a Qiagen Genomic column (Genomic tip 100) and precipitate the DNA as recommended by the manufacturer, resuspending the DNA in 50–500 μl TE buffer.

Although supercoiled DNA has been preferred in transformation, linear DNA such as DNAs generated from PCR reactions can be directly injected alone or with supercoiled co-injection markers. The rate of transformants and the fate of injected DNA appear to be similar to that obtained with super-coiled DNAs (P. Anderson and L. Avery, personal communications). Transformation to disrupt gene function can also be performed using RNA, as described in Chapter 6 by R. Barstead.

4.1.2 Making an injection mixture

The concentration of DNA in the injection mixture depends largely on the specific DNA sequences and constructs. A standard injection mixture should include about 10–100 μg/ml of tester DNA and 20–100 μg/ml co-injection DNA. The total DNA concentration in the injection mixture is recommended to be more than 100 μg/ml in order to form a stably transmitted extra-chromosomal array (5). Thus, if a high concentration of tester DNA is toxic to the worm, one may compensate by raising the amount of co-injection DNA or supplementing it with inert DNA sequences such as DNA size ladders.

An injection buffer (final concentration: 20 mM KPO_4, 3 mM K citrate, 2% polyethylene glycol (PEG) 6000 (pH 7.5)) was recommended for injection into nuclei (2). However, this solution is rather viscous and can clog the needle. It is not essential for successful transformation.

4.1.3 Preparing a clean injection mixture

All DNA preparations contain some forms of contamination that can cause needle clogging. It is important to remove this contamination by centrifuging the DNA as recommended in *Protocol 4*.

Protocol 4.   Making a clean injection mixture

Equipment and reagents

- 10 × injection buffer (if used): 200 mM KPO_4, 30 mM K citrate, 20% polyethylene glycol (PEG) 6000 pH 7.5
- DNAs to be used in microinjection, i.e. DNA of interest and co-injection marker DNA

Method

1. Before taking DNA from the stock, centrifuge the DNA solutions in a microcentrifuge at 20 000 g (14 000 r.p.m.) for 2–5 min, and use the top few microlitres to prepare the injection mixture.

2. Prepare 15–20 μl of injection mixture containing the DNA of interest and co-injection marker DNA at desired concentrations, maintaining the total DNA concentration at more than 100 μg/ml. Make up the rest of the volume with distilled water. (Injection buffer (see text) may be used.)

3. Centrifuge the injection mixture in a microcentrifuge at 20 000 g (14 000 r.p.m.) for 10–15 min and carefully transfer the top 2–3 μl to a new tube for immediate use in injection. If the injection mixture is not finished that day, recombine the mixture into one tube and centrifuge again before use.

4. Break step 3 into two centrifugations if the DNA mix still causes clogging problems after this procedure. After the first centrifugation, transfer the top 10 μl to a new tube, centrifuge for 5 min, and use the top 2 μl after the second centrifugation for injection.)

4.2 Needle preparation and sample loading

4.2.1 Pulling a good needle

Microinjection needles are made from fine, glass micropipettes of diameter 1.0–1.2 mm. Micropipettes are sold by many electrophysiology suppliers; e.g. World Precision Instruments Inc.(Sarasota, FL, USA) and Clark Electro-medical Instruments (http://www.clark.mcmail.com/). We use borosilicate glass capillaries from World Precision Instruments (cat. no. 1B100F-4).

The quality of a needle is the key to a successful transformation. Various brands of needle pullers are operated by different programs that control several parameters, including the coil temperature, the heating time, the pulling force, and speed. No matter which needle puller is used, one must calibrate the machine to find the optimal settings by trial and error. The key parameter appears to be the coil temperature. For a crude assessment of the needle's quality, examine the tip of the needle under 40–50 $\times$ magnification on a dissecting microscope: the needle tip should taper quickly and smoothly. If the needle is already open, it is unlikely to be useful because the needle is probably too wide and blunt. If the needle has a stubby taper, the needle may be good but may not leave enough space for breaking the tip (see later). If the needle has a long thin taper, the needle may not be stiff enough. Finally, one can put the needle into the needle holder and move the needle back and forth in the injection oil; a good needle will not wobble. Once a good setting is found, one can pull several needles and save them in clean containers such as a Petri plate, with clay or double-sided tape to secure the needles and prevent accidental damage of the needle tips.

4.2.2 Loading the injection mixture

Sample loading is by capillary action. One can load the injection mixture into the needle hub either using a hand-pulled, mouth pipette or a standard Pipetman tip. To pull a mouth pipette, heat the centre of a capillary (50–200 μl) over a flame, and as it softens gently pull the capillary apart. Typically, a 0.2–0.5 μl solution is used per loading, which should allow the injection of more than 100 worms. To avoid air bubbles accumulating in the needle or to prevent dirt flowing towards the tip, one can hold the loaded needle in an upside down position for a minute after loading. Loaded needles should not be stored in the open air for more than a day. Once loaded, needles can not be reused for different DNA samples.

4.2.3 Breaking the tip of the needle

Most needle pullers produce sealed ends that need to be broken open. This is frequently done by rubbing the end of the tip against a thin glass bar made by pulling a heated microcapillary. Liquid flowing out of the needle indicates that the needle is open. The quality of the needle opening is judged by both the shape of the opening and the flow rate. Moving the tip of the needle towards the glass bar at an angle helps prevent the generation of a blunt opening. Opening the tip under the 40 × objective allows one to observe directly how much of the needle has been broken. By doing so, one can also get a clear idea of whether the needle is stiff enough. A good needle will break easily, whereas a soft and thin needle will require more effort, often an indication that the needle may not be usable. To judge the flow rate from the needle, set the injection pressure around 15 p.s.i. and one should see a single drop of liquid leaking from the needle under the 5–10 × optics. However, a thinner but good-shaped and strong needle can be used by increasing the injection pressure, (occasionally one can complete the injection by using the tank pressure); a needle with a wide but sharp opening can be used by lowering the injection pressure.

4.2.4 Clearing the pathway through the needle

Often small air bubbles remain stuck at the tip of the needle. These may disappear by simply leaving the needle under the resting pressure for a few minutes. If the needle is broken with a proper opening size, the air bubbles can be easily expelled by applying the tank pressure briefly. If air bubbles stay inside the needle under tank pressure this indicates that the opening is either too small or that it is clogged by contaminants in the injection mixture.

After needles are adjusted as desired, align the needle in a central position to reduce the manoeuvring time of alignment during injection. *The needle should always be lifted up from the focal plane for injection when not in use to avoid accidentally impaling injected worms and/or damaging the needle tip.*

4.3 Mounting worms

4.3.1 Preparing host worms

Although any adult worms can be used for injections, well-fed, clean, and healthy young hermaphrodites are best. The gonad of a healthy worm has the characteristic morphology shown in *Figure 2*, which is easy to recognize. The cuticles of healthy worms are strong and elastic, allowing good recovery after injection. To keep healthy worms, pick 20–50, well-fed, L4 host worms on to a fresh seeded plate the day before injection.

4.3.2 Preparing agarose pads

Worms are immobilized for injection on dried agarose pads. *Protocol 5* describes how to make the pads. Pads can be stored in a clean container for a long time. Before use, the pad should be moisturized by breathing briefly over the surface.

Protocol 5. Making agarose pads for injections

Equipment and reagents
- Regular gelling-temperature agarose
- Glass coverslips (22 × 50 mm or 25 × 50 mm)

Method

1. Prepare 5 ml of 2% regular gelling-temperature agarose in water, and melt the agarose completely.

2. Using a Pasteur pipette, add a drop of melted agarose on to a cover-slip, immediately put another coverslip on top of the agarose, and press gently on the top coverslip to flatten the agarose. Prepare 5–10 more pairs of coverslips in this way.

3. Separate the two coverslips either by lifting the top one or sliding the two apart.

4. Leave the agarose pad in an 80°C drying oven for an hour or longer. Alternatively, air-dry the pads overnight at room temperature or at 37°C for an hour or longer.

5. Save the remaining 2% agarose for up to several days, but cover the opening and add an appropriate amount of water to regenerate the starting agarose concentration.

4.3.3 Mounting worms

The worms are immobilized on to the dried agarose pad with the dorsoventral axis of the worm parallel to the surface of the coverslip (see *Protocol 6*). When mounted properly, the entire gonad arm should be seen in one focal plane. This step is technically challenging and is definitely the most time-consuming. Beginners should only mount a single worm per agarose pad because worms quickly dessicate on the pad. Dessicated worms appear shrunken and lack contrast. When mounting multiple worms on the pad, keep them all in the same orientation to save time in alignment during injection.

Protocol 6.   Mounting worms on agarose pads

Equipment and reagents

- Agarose pads
- Halocarbon oil series 700 (Halocarbon Products Corp., cat. no. 9002–83–9)
- Healthy young adult hermaphrodite worms
- Wormpick and dissecting microscope

Method

1. Moisturize the dried agarose pad by briefly breathing over the surface. This also makes it obvious which side of the coverslip the agarose pad is fixed to.

2. Add a drop of halocarbon oil to the pad. Pick one or a few worms and place them into the oil under the microscope.

3. Let the worm thrash about in the oil to remove attached bacteria. The viscosity of the oil also slows the movement of the worm.

Protocol 6. *Continued*

4. Using a wormpick, gently force the worm on to the pad while maintaining the dorsal and ventral sides of the worm in the same focal plane.

5. Once the worm is close to the surface of the pad, press firmly and gently to stick the worm on to the pad: have part of the worm body stuck first, then press along the rest of the body so that the entire worm sticks to the pad. (Do not stretch the worm.)

4.4 Injections

4.4.1 Aligning the needle with the worms

Once the worms are mounted, transfer the coverslip immediately to the injection scope to align the needle with the worm.

(a) Centre the worm under low magnification.

(b) Bring the needle to the same focal plane swiftly and carefully by lowering the needle holder.

(c) Switch to high magnification and use the fine controls of the manipulator to bring the tip of the needle against the worm cuticle adjacent to the syncytial gonad.

The needle should be inclined at an angle of 10–20°. To speed up the alignment and avoid extensive stage-rotation, one should remember how the worms are arranged from when they were viewed under the dissecting microscope.

4.4.2 Injecting into a gonad

For an excellent needle, touching firmly against the cuticle of the worm will allow the needle to penetrate the cuticle to reach the gonad. However, usually a gentle tap on the micromanipulator, while the tip of the needle is firmly against the cuticle, is needed. Applying pressure to the DNA solution allows it to flow into the gonad (see *Figure 3*). When the gonad is completely filled and appears as an inflated sausage-shaped balloon or when excess liquid flows out of the gonad, end the injection by removing the needle from the worm. If using a manipulator with fine control for the horizontal axis, withdraw the needle from the gonad with the manipulator. If using a manipulator with fine control only along the vertical axis, pull the worm away from the needle by moving the gliding stage. If possible, both gonad arms should be injected. The needle should be maintained in a central position if multiple worms are to be injected for the same pad.

4.4.3 Protecting and reshaping the needle during injection

Needles are easily clogged by cytoplasm, cuticle, or bacteria around the worm. To protect the needle from clogging, remove the needle from the worm

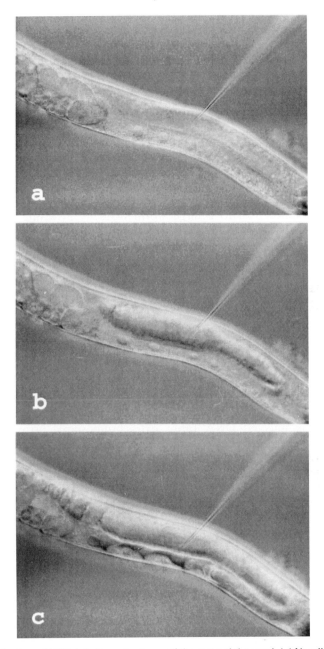

Figure 3. Injection of DNA into the cytoplasm of the syncytial gonad. (a) Needle is inserted into the gonad at the widest point in the gonad syncytium. (b) Fluid is expelled from the needle with moderate pressure; the gonad swells up around the point of injection. (c) Fluid fills up the whole gonad, which now acquires a sausage-like appearance. The needle is withdrawn before the gonad bursts. (Reproduced, with permission, from ref. 5.)

while under injection pressure. This may seem a waste of injection mixture, but in fact it helps to keep the worm alive (see later). Between each injection, use the high pressure to clear the tip of the needle. Clogged needles can be used after re-breaking the tip against the glass bar.

4.5 Recovery of injected worms

It is important to recover worms from the dry agarose pad as soon as possible. Add a single drop of M9 buffer on to the surface of the oil and bring the M9 buffer in to contact with the worms using a wormpick so that the worms float away from the agarose pad. Immediate wriggling of the worm indicates a good recovery. The worms can then be transferred on to a fresh seeded plate using a wormpick, a mouth pipette, or by simply dropping the worms (along with the M9 buffer) on to a large worm plate by inverting the injection coverslip.

A recovery buffer (final concentration: 0.1% salmon sperm DNA, 4% glucose, 2.4 mM KCl, 66 mM NaCl, 3 mM $CaCl_2$, 3 mM Hepes pH 7.2) is recommended in place of M9 buffer for recovering worms that are fairly desiccated (2, 6). However, it is simpler and easier to aim to recover all injected worms rather than attempting to inject too many worms at one time.

4.6 Identification of transformants and establishing lines

Transformants are identified mostly by co-injection markers that confer either an obvious visible phenotype, such as a dominant behaviour or rescue of a mutant phenotype or detectable expression from a reporter gene.

Because the injected DNA forms extrachromosomal arrays of various structures, one should keep a careful record of independent transformants. Since only a fraction of F_1 transformants produce heritable arrays, one to five F_1 transformants (depending on how many are obtained per injection) can be placed per seeded plate. *Only one line from each plate should be kept and treated as an independent line.* Depending on the host strains and co-injection markers, some transformant lines need to be maintained by picking transformants every generation, whereas others need less maintenance because transformants have a growth advantage over non-transformants.

4.6.1 A guideline for the self-assessment of injection skills

Learning the technique of transformation in *C. elegans* may take a few hours or several days of persistent practice. With healthy host worms that have normal brood sizes, a good injector should be able to:

- inject about 20–40 hermaphrodites (P_0) in an hour;
- obtain about 10–20 F_1 transformants per injected P_0;
- at least 10–50% F_1s should produce germline transmitted F_2s.

4.7 Troubleshooting

Transformation in C. elegans is a difficult technique. Skills in manipulation of worms and needles are learnt with practice. Each of the steps described above can be practised separately. One should perform initial injections using healthy host worms and DNAs that are not toxic to the worm. Beginners may need to overcome their frustration with the technical difficulties, but they should be confident of mastering the technique. Perseverance usually leads to success in generating transformants. For laboratories that have no prior experience with C. elegans, a visit to a nearby C. elegans laboratory would be valuable. A video monitor attached to the microscope is very useful for teaching the injection technique.

4.7.1 Common problems

(a) *If needles keep clogging after breaking them open*:
   (i) The injection mixture is not clean enough. Remake the injection mixture, centrifuge for longer, and add an additional centrifugation step, taking only the top of the solution for injection.
   (ii) The glass bar for breaking the needle is too dirty. Replace it.
   (iii) Avoid clogging the needle with cytoplasm by pulling the needle out of the gonad while the injection pressure is on and clearing the needle by applying the tank pressure between injections.

(b) *If needles do not break*:
   (i) The needle tips may be too thin. Lowering the coil temperature when pulling the needles may help.

(c) *If needles are blunt and too wide*:
   (i) The needles may taper too abruptly. Raising the coil temperature when pulling the needles may help.
   (ii) The glass bar may be too thick.
   (iii) You may not be breaking off the needle tip gently enough. Watch the flow rate from the needle while breaking the tip under 40 × magnification.

(d) *If worms cannot be loosened from the pick*:
   (i) Use injection oil, instead of bacteria, to pick the worms off the plate.

(e) *If worms do not stick to the agarose pad*:
   (i) Bake the pad at 80°C and moisturize the pad before use.
   (ii) There may be too many bacteria around the worm. Clean the worm-pick by flaming it, and having the worm thrashing about in the oil helps to remove excess bacteria.
   (iii) Stick one part of the worm body to the pad first, then use an injection needle to push the rest of the worm on to the pad. This gives better control and allows one to see how close the worm is to the pad.

(f) *If it is difficult to get the needle into the worm*:

    (i) Needles may be too blunt. Prepare the needles again.

    (ii) Worms move too much because of inadequate mounting.

    (iii) Worms are dying.

    (iv) Needles are broken too wide so that liquid flows from the needle upon touching the worm, resulting in the worm floating off the agarose pad. Reduce the injection pressure to stop the liquid flow.

(g) *If worms do not recover*:

    (i) You may have damaged worms at any of the preceding steps.

    (ii) Worms are desiccated because it took too long to stick them on to the pad. Try to work with one worm at a time. Adding an extra drop of oil on to the worms after mounting may help prevent drying. Releasing some injection mixture around the worm helps it to survive longer. This is particularly helpful if multiple worms are injected from the same pad.

    (iii) Worms may die, before injection takes place, because they are over-stretched during mounting. Try to maintain the body curvature of the worm while pressing it on to the pad. The worms that recover best are those that still wriggle their heads and tails during injections.

    (iv) Put all injected worms on to a recovery plate and transfer the viable ones to new plates to reduce the amount of plates wasted.

(h) *If no lines are seen after worms survive all procedures*:

    (i) You may have injected the DNA into the wrong place. If the DNA is injected into the gonad, you should see the gonad fill up and the liquid flow slow down. If the DNA injected flows quickly into the worm, you may have injected into the gut. The gonad has a limited capacity, where as the gut apparently doesn't.

    (ii) You may damage the cells in the gonad. Score the brood size of injected worms, and make sure that it is nearly the same as for uninjected worms.

(i) *If no lines are found after perfect injections*:

    (i) The host worms are sick, have abnormal gonads, or have small brood sizes. Using other worm strains as hosts may avoid the problem.

    (ii) The DNA is toxic to worms, either inherently due to the sequence or because of the mode of preparation. Repetitive sequences can be toxic, but it is not clear why some are toxic while others are not. In most cases, toxicity is not due to the coding sequences of the gene of interest, but to intronic sequences or sequences in between genes. To circumvent this problem, one can lower the DNA concentration, or select DNA fragments that overlap with the essential sequences of the DNA fragment of interest, but lack non-essential 5' or 3' sequences. If you have switched to using a new DNA purification method, test by purifying a DNA known to be non-toxic using the same procedure.

5. Fate of injected DNA

It is unfortunate that the fate of DNA injected into the worm is not well understood. Most of what we know comes from Southern blot analysis that has limited resolution. The fate of injected DNA depends on several factors: site of injection, co-injection markers, and the sequence of the DNA. Stinchcomb et al. (3) first described the formation of long tandem arrays of a plasmid injected into the syncytial gonad. The plasmid appears to be linearized randomly, and several hundred copies of the plasmid are ligated to form one array. The arrays are heritable and are segregated as extrachromosomal elements at frequencies varying between 5 and 95%. On the other hand, after injecting DNA containing the amber suppressor sup-7 DNA into the oocyte nuclei, Fire (2) obtained both integrated and non-integrated arrays that are composed of a low copy number (1–10) of the injected DNA. Integration occurs in a non-homologous fashion. Formation of low copy number arrays in this instance may be because high levels of amber suppressor tRNA are deleterious to the animal.

The fate of injected DNA has been summarized by Mello and Fire (6), based on many published and unpublished observations. A brief summary of which follows. The injected DNA is frequently expressed in a mosaic fashion in the first generation (F_1) following microinjection, such that only a fraction of cells in a given tissue express the DNA. This F_1 mosaic expression appears to result from simple partitioning of the injected DNA among the cells of the F_1 embryo. In a fraction of F_1 progeny, the injected DNA forms arrays in which the injected DNA is arranged in tandem repeats. The formation of an array may be the result of recombination and ligation reactions driven by homologous recombination between injected DNAs or other unknown factors. Co-injected DNAs are usually mixed in an array, in random order and in proportions that are the same as the relative DNA concentrations in the injection mixture. However, if one of the injected DNAs is toxic to the worm when present in high copy numbers, an array may contain an odd ratio of co-injected DNAs or exclude the toxic DNA. Arrays greater than 700 kb are transmitted through the germline as extrachromosomal elements, whereas smaller arrays are lost unless integration into a chromosome occurs. The extrachromosomal arrays behave like chromosomal free duplications in that they display non-Mendelian inheritance. The inheritance of independently established arrays ranges from 10–90%. These differences in heritability may reflect a simple size difference between arrays: larger molecules may be replicated or segregated more efficiently than smaller ones. Thus, *independently established arrays must be treated as distinct genetic entities* (see above). Spontaneous integration of the arrays after injection into gonad syncytia is rare, whereas integration into a chromosome occurs at moderate frequency if the DNA is injected directly into the oocyte nucleus. Homologous integration has been observed serendipitously (ref. 11 and A. Fire, personal communication), but our understanding of this phenomenon is poor.

Even less is known about expression from the injected DNA in a given array. There has been no systematic analysis of correlations between the copy numbers of an injected DNA versus levels of RNA and protein products. In several integrated arrays constructed using standard concentrations of *egl-10(+)* and *rol-6* as the co-injection marker, a 20-fold excess expression of EGL-10 protein, as compared to that in wild-type worms, was detected on Western blots—along with EGL-10 cross-reactive proteins, of various sizes, that are likely to be EGL-10 fusion proteins resulting from random recombination during array formation (M. Koelle, personal communication). Moreover, while the structure of the extrachromosomal arrays seems stable, the expression levels of genes from extrachromosomal arrays change when maintained for many generations. Thus it is important to make frozen stocks as soon as an array is generated to ensure consistency during a long period of analysis.

6. Integration of extrachromosomal arrays

Extrachromosomal arrays can be stabilized by a mutagenesis-induced integration into a chromosome. Integration of an array may alleviate mosaic expression, and also provides convenience in standard genetic crosses and screens. It can be done by any conventional mutagenesis technique, although most popular procedures use mutagens that introduce chromosomal breaks. When an X-ray or gamma-ray machine is not available, one can use psoralen or EMS (S. Shaham, personal communication) as mutagens. There are numerous ways to identify integrants, largely depending on how the array is created and how eager one is to obtain the integrants. *Protocol 7* describes a procedure used in my laboratory.

Protocol 7. X-ray or gamma-ray induced integration of extrachromosomal arrays

Equipment and reagents

- Two or three independently generated array-bearing strains. Select strains with arrays that transmit at a relatively low rate, less than 50%.
- X-ray or gamma-ray source
- Seeded NGM agar plates (see *Protocol 2* and Section 3.3 in Chapter 4)

Method

1. Pick 20–30, well-fed, L4 worms containing the arrays on to a fresh plate. Irradiate the plate in an X-ray machine or gamma-ray machine with 3000–4200 rads. Since each machine is calibrated differently, irradiate two batches of worms at different doses for each strain. The proper irradiation dose should result in 10–15% lethality and sterility in F_1 progeny.

---

2. Transfer three or four irradiated worms (P_0) on to each of six to eight separate plates.

3. From the F_1 progeny, transfer 300–500 array bearing individual worms to individual plates. To avoid isolating the same integrative event many times, pick the worms from separate P_0 plates and keep a record of the origin of each worm.

4. Pick two to four individual F_2 transformants for each F_1 on to separate plates. Examine the F_3 progeny for 100% transmission of the transformation marker.

Note: It is often an indication of successful integration if more than one F_2 plate gives 100% F_3 transformants. If the integrants observed are from different P_0 plates, the integrations are independent.

The mechanism of integration has not been well characterized. The entire extrachromosomal array appears to be incorporated into the genome as a single element. Insertions occur largely in a random fashion; however, hot spots in the genome for insertion of specific arrays have been observed in several cases (Y. J., unpublished results, and E. Jorgensen, personal communication). Insertion of an array may also affect chromosomal behaviour, such as suppression of chromosomal recombination (J. Way, personal communication).

There are various effects of integration on gene expression from the array: expression may be higher or lower than that from the original extrachromosomal array; in some cases, after integration, the expression from the DNA tested is completely abolished, with no effect on the co-injection markers. This may be due to positional effects caused by disruption of chromatin structures. Sometimes a mutant phenotype is found associated with the integration, which may or may not relate to the expression of the gene of interest on the array. All integrants should be backcrossed several times to remove other mutations caused by the mutagenesis. Integrants are not always stable, loss of apparently integrated arrays after passage for many generations has been observed (Y. J., unpublished results). Such losses might be explained if the array was attached to the end of, rather than inserted into, a chromosome. How this might occur is unknown.

When using integrants to analyse gene expression, multiple integrants of the same DNA composition, but from independent arrays, should be obtained. First, integration does not always solve the problem of mosaicism. The degree of mosaicism apparently depends on many variables such as the promoters used in expression, the reporter genes, and the copy numbers of the transformed DNA, etc. Second, the expression pattern from the integrated array may be modified by nearby irrelevant enhancer or suppressor elements. Hence, extrapolating data from several independent integrants may avoid misinterpretations caused by positional effects, which are likely to be due to changes in chromatin structure or fortuitous enhancer elements near the insertion site.

7. Transformation markers

Co-injection markers are used routinely to identify transformants. It was thought that for two different DNAs to be included in one array, they must share some sequence homology (5). However, reports that PCR-generated linear DNAs and supercoiled co-injection markers that share no homology have been found in the same array (W. Davies and L. Avery, personal communications) undermines the stringent requirement for homologous sequences.

The choice of a co-injection marker depends critically on the experimental design. However, the co-injection marker should allow for the easy identification of transformants and should result in minimal interference with the expression of the DNA tested. Since nearly all co-injection markers are found to cause some improper expression of co-injected genes, it is important to include proper controls.

Traditionally, genes, either wild-type or those containing a mutation, that confer a recognizable phenotype upon expression are used as co-injection markers. To identify transformants, appropriate host worms may need to be constructed. Recently, because of concerns about complications from over-expression of the gene under study, reporter gene constructs in which *GFP* or *lacZ* is under the control of a *C. elegans* promoter have been used. These reporter markers can be detected in any genetic background, eliminating unnecessary work in constructing strains. Another advantage of reporter genes is that one can avoid potential interference from co-injection markers by choosing a construct in which the expression of the reporter gene shows no overlap with that of the gene of interest. Although *lacZ* is still used in certain cases, *GFP* has become the most popular choice except in the study of early embryos. If *GFP* is used as the co-transformation marker, a fluorescence dissecting microscope (Leica; Kramer Scientific) may be necessary for identifying transformants. Some co-injection markers that have been used are given below.

7.1 Wild-type or mutant genes as co-injection markers

(a) *rol-6 (su1006)*: The plasmid, pRF4 (12), can be obtained from J. Kramer (Chicago, IL). This marker has been popular because the phenotype is readily identified under 5 × magnification. The dominant mutation in *rol-6(su1006)* results in a right-hand roller phenotype. However, the Rol phenotype can only be detected in the L2 larval stage and onward. Thus it is not useful for identifying L1 and embryonic transformants. This marker may not be used in certain genetic backgrounds, such as animals that roll

anyway and animals that suppress the Rol phenotype. The Roller phenotype makes it difficult to identify many cells in the transformants. Roller
males do not mate efficiently, making genetic crosses difficult. Because
Rol animals often have a growth disadvantage over non-Rol animals,
lines must be maintained by successively picking Roller animals. This
marker can result in suppression of expression in the nervous system (C.
Bargman, personal communication) and enhancement of expression in
the pharynx (A. Fire, personal communication).

(b) *sup-7(st5)*: The plasmids, pPD26.14 or pAST (13), can be obtained from
A. Fire (Baltimore, MD) and contain an amber suppressor mutation in
the *sup-7* tRNA gene. Strains containing the *tra-3 (e1107)* amber allele
are frequently used as the host. The presence of *tra-3* is required to
specify the hermaphrodite fate (14). Hermaphrodites carrying *tra-3 (e1107)*
are partially masculinized (pseudomales). Homozygous *tra-3* animals can
not be propagated at temperatures above 20 °C because they are sterile.
The Tra-3 phenotype can be maternally rescuable, thus expression of
suppressor tRNA in either mother or zygote can result in animals that are
fertile at temperatures above 20 °C. The advantage of this marker is that
copy numbers of injected DNA are limited due to toxicity of high levels
of amber suppressor tRNA. It also appears to result in a higher frequency
of integrative transformation (2, 11). Whether this is a direct effect of
injection into oocyte nuclei or a combination of the marker plus the
injection site remains unknown.

(c) *unc-22* antisense DNA: The plasmid, pPD10.46, can be obtained from A.
Fire (Baltimore, MD). *unc-22* encodes a muscle filament protein, twitchin,
that functions in muscle contraction (15). Expression from this DNA
construct produces *unc-22* antisense RNA that interferes with normal
UNC-22 function, resulting in a loss of *unc-22* function phenotype:
animals have low mobility, twitch, and are egg-laying defective. The
twitching phenotype is more prominent in 1 mM nicotine. The advantage
of this marker is that the phenotype can be seen in all larval and adult
stages, and cell identification is easy. The drawbacks of this marker are
that the phenotype is relatively subtle, and it is not useful for paralysed
host worms. Similarly, twitching males have a low mating efficiency. In
addition, the plasmid appears to contain a strong specific enhancer for
body-wall muscle expression. Because transformants are less healthy than
non-transformed hosts, lines must be maintained every generation by
picking Unc worms.

(d) *dpy-20(+)*: The plasmid, pMH86 (16), can be obtained from M. Han
(Boulder, CO). The *dpy-20* marker encodes a novel protein that may be
involved in cuticle formation (17). It has been used as a co-injection
marker where cell identification is important and *rol-6(su1006)* is inappropriate. Host strains contain *dpy-20(e1282ts)*, which causes a medium

dumpy phenotype at 22.5 °C. Rescued animals are wild-type in size or long animals that may result from overexpression of *dpy-20(+)*. This marker appears to have complex enhancer and suppressor effects on co-injected DNA (Y. J., unpublished results; S. Siddiqi, personal communication).

(e) *pha-1(+)*: The plasmids, pBX or pC1, can be obtained from R. Schnabel (Braunschweig, Germany). *pha-1* encodes a bZip-like transcription factor expressed in the pharynx cells (18). The host animals are *pha-1(e2123ts)* that grow normally at 15–20 °C, but are 100% embryonic lethal at 25 °C. After injection, transformants can be selected by raising them at 25 °C, thus any survivors would contain the introduced DNA. This marker can reduce labour in the maintenance of transformant lines and can allow large numbers of transgenic worms to be obtained for large-scale RNA or protein purifications. However, the effect of *pha-1* on the expression of co-injected DNA has not been analysed.

(f) *lin-15(+)*: The plasmid, plin-15EK (17), can be obtained from R. Horvitz (Cambridge, MA). *lin-15* encodes two novel proteins that function in the hypodermis in vulva formation. The host is *lin-15(n765ts)*: animals at 15 °C are wild-type, but multivulva (Muv) at 22.5 °C or higher temperatures. Because Muv animals are sicker than wild-type transformants, there is less labour in maintaining the lines. It has been favoured by many people working on the nervous system. However, some reports suggest that overexpression of *lin-15(+)* may cause low-penetrance, axonal path-finding defects and suppression of expression in certain neurons such as the HSN (G. Garriga, personal communication). Other reports suggest that the Muv phenotype is not always fully rescued by expression of this plasmid (R. Baran, personal communication).

A few less commonly used markers include: *unc-4*, which functions in the DA and VA ventral cord motor neurons to rescue a backward Unc phenotype (19), and the plasmid can be obtained from D. Miller (Nashville, TN). *unc-76* rescues a severe Unc phenotype (20) and the plasmid can be obtained from R. Horvitz (Cambridge, MA); *unc-30* functions in the DD and VD ventral cord motor neurons to rescue a shrinker phenotype (21), and the plasmid can be obtained from Y. Jin (Santa Cruz, CA).

7.2 Reporter constructs as co-injection markers

(a) *Actin::lacZ*, a translational fusion of the *C. elegans act-4* gene and *E. coli lacZ* (22), can be obtained from J. Shaw (St Paul, MN). *LacZ* expression is limited to several tissues including body-wall muscles, vulval muscles, spermatheca, and the pharynx. The eggs of animals stained with X-Gal remain viable, allowing the recovery of transformants. This marker has not been widely used, but should be considered when GFP is not appropriate.

(b) *unc-119::GFP* is expressed in all neurons (23), and can be obtained from D. Pilgrim (Alberta, Canada). It produces bright fluorescence that is seen easily in all stages and is particularly good for identifying embryonic transformants.

(c) *sur-5::GFP* is expressed in the nuclei of all somatic cells, can be obtained from M. Han (Boulder, CO), and is recommended for mosaic analysis (see later) (24).

(d) *egl-5::GFP* is expressed in many cells in the posterior embryo and can be obtained from A. Chisholm (Santa Cruz, CA).

(e) *ceh-19::GFP* is expressed in a few amphid neurons, can be obtained from Y. Jin (Santa Cruz, CA) (25), and can be very useful as a transformation marker for analysis aimed at the mid-body region of the worm.

(f) *glr-1::GFP* is expressed in many interneurons (26), and can be obtained from C. Bargmann (San Francisco, CA).

8. Gene cloning by germline transformation using the wild-type DNA

Transformation is essential for many types of gene manipulation in *C. elegans*. In this section, I discuss general guidelines on the application of transformation in gene cloning. For the use of transformation in analyzing gene expression patterns, overexpression and/or ectopic expression of genes of interest, see Chapter 9. For the use of transformation in mosaic analysis, see Chapter 12.

A standard procedure for gene cloning based on rescuing a mutant phenotype follows. First, map the gene of interest by standard genetic methods (two-factor, three-factor mapping, or mapping by deficiency and duplications (see Chapter 12)) in relation to landmarks, such as cloned genes, which connect the physical and genetic maps; other mapping methods using restriction fragment length polymorphisms (RFLP) or sequence tagged sites (STS) can be used in conjunction to further refine the region of interest (27). Second, when the region of interest is represented by less than 10–50 cosmid clones, one can attempt to rescue the mutant phenotype by germline transformation using pools of overlapping cosmid clones. There is no limit to the numbers of clones that may be pooled together in an injection mixture since the different clones are recombined independently in formation of an array. DNA pools covering continuous *C. elegans* genomic sequences of up to 0.5 Mb have been used for injection (28). However, the more complex the pool is, the higher the chance of missing a rescuing event. Third, when a rescuing activity is found associated with a specific pool, successive germline transformations (using individual cosmids from the pool and subclones from the rescuing cosmid) can then be performed to identify the gene. Fourth, the identity of the gene is confirmed by finding DNA sequence alterations in mutant alleles or by disrupting the integrity of the rescuing fragment.

Before pursuing germline transformation, one should know several facts about the gene of interest: Is the mutant host worm rescuable by addition of a wild-type copy of the gene? Will the mutant phenotype of the host be changed by the co-injection marker? How will the rescue be scored? Simple recessive loss-of-function mutations are rescuable by adding wild-type copies of the gene. Gain-of-function mutations may or may not be rescuable depending on the nature of the mutation; increasing the expression of the wild-type gene should not rescue hypermorphic mutations, but may rescue neomorphic mutations. One should examine the effects of a wild-type gene on the mutation by standard dosage analysis. The effect of co-injection markers can be assessed by generating transformants containing only the co-injection marker. If mutants display multiple defects, one should score all phenotypic defects for a complete rescue. However, because transgenes may not be expressed in all tissues, partial rescue of some phenotypic defects may be the best indication of cloning the correct gene (see later). For rescuing a mutant phenotype that is incompletely penetrant, a large amount of data must be collected on genotypically comparable worms.

When to start cloning a gene by germline transformation depends on balancing the effort of genetic mapping and microinjection. The finer the definition of the genetic map position of the gene, the less microinjection one needs to do. If possible, when choosing the cosmids to use, inspect the genomic DNA sequence and avoid using cosmids in which a candidate gene is close to the end of the cosmid; the cosmid may lack essential 3' or 5' sequences. Because of possible toxicity associated with some cosmid DNAs, start with a relatively low concentration of cosmid DNA, such as 1–10 μg/ml. Moreover, because the formation of an extrachromosomal array is a rather random process, there can be several reasons for failing to observe transformation rescue after testing all possible cosmid DNAs. First, when the complexity of the injected DNA is high, not all cosmids in the injection pool are included in a single array. Second, even when a cosmid of interest is in the array, the gene of interest may not be intact after array formation. Third, there may be insufficient expression of the gene of interest because the expression level of a gene from an array is unpredictable. Thus when using multiple cosmid DNAs in an injection mixture, several (as many as 10) independent arrays should be generated before drawing any conclusions. This principle is also true for YAC DNA, which is often 200 kb or longer and sheared into smaller fragments during preparation of the injection mixture.

Not all mutants can be rescued by transformation with the wild-type DNA, e.g. genes that function specifically in the germline. Until now, transgene expression in the germline has been unsuccessful. The germline appears to have specific mechanisms to silence transgenes (29). This could be specific transcriptional repression of the transgene or chromatin incompatibility between transgenes and the germline transcriptional machinery. By including a large amount of random *C. elegans* genomic DNA in the injection mixture, Kelly *et al.* were able to observe germline expression of reporter genes driven by the *let-858* promoter, a gene that is widely expressed in both soma and germline (29). An alternative method for cloning genes that function in the germline is through RNA interference to phenocopy the mutant phenotype, as described in Chapter 6.

Acknowledgements

I thank the many people who sent their protocols and provided invaluable unpublished information, in particular M. Zhen, A. Davies, J. Shaw, C. Bargmann, W. Davis, C. Mello, M. Han, A. Chisholm, M. Koelle, G. Garriga, R. Baran, A. Fire, and members of the Jin and Chisholm laboratories at UCSC. Many thanks to A. Chisholm, A. Zahler, and I. Hope for their comments on the manuscript.

References

1. Kimble, J., Hodgkin, J., Smith, T., and Smith, J. (1982). *Nature* **299**, 456.
2. Fire, A. (1986). *EMBO J.* **5**, 2673.
3. Stinchcomb, D. T., Shaw, J. E., Carr, S. H., and Hirsh, D. (1985). *Mol. Cell. Biol.* **5**, 3484.
4. Fire, A. and Waterston, R. H. (1989). *EMBO J.* **8**, 3419.
5. Mello, C. C., Kramer, J. M., Stinchcomb, D., and Ambros, V. (1992). *EMBO J.* **10**, 3959.
6. Mello, C. and Fire, A. (1995). In *Methods in cell biology*: Caenorhabditis elegans: *modern biological analysis of an organism* (ed. H. F. Epstein and D. C. Shakes), p. 451. Academic Press, San Diego, CA.
7. Kimble, J. and Ward, S. (1988). In *The nematode* Caenorhabditis elegans (ed. W. B. Wood), p. 191. Cold Spring Harbor Laboratory Press, NY.
8. Schedl, T. (1997). In C. elegans *II* (ed. D. L. Riddle, T. Blumenthal, B. J. Meyer, and J. R. Priess), p. 241. Cold Spring Harbor Laboratory Press, NY.
9. Gaugler, R. and Hashmi, S. (1996). In *Genetic engineering*, Vol. 18 (ed. J. K. Setlow), p. 135. Plenum Press, New York.
10. Sambrook, J., Fritsch, E. F., and Maniatis, T. (1989). *Molecular cloning: a laboratory manual.* Cold Spring Harbor Laboratory Press, NY.
11. Broverman, S., MacMorris, M., and Blumenthal, T. (1993). *Proc. Natl Acad. Sci. USA* **90**, 4359.
12. Kramer, J. M., French, R. P., Park, E.-C., and Johnson, J. J. (1990). *Mol. Cell. Biol.* **10**, 2081.
13. Fire, A., Kondo, K., and Waterston, R. (1990). *Nucleic Acids Res.* **18**, 4269.
14. Hodgkin, J. A. and Brenner, S. (1977). *Genetics* **86**, 275.
15. Moerman, D., Benian, G., Barstead, R., Schriefer, L., and Waterston, R. (1988). *Genes Dev.* **2**, 93.
16. Han, M. and Sternberg, P. W. (1991). *Genes Dev.* **5**, 2188.

17. Clark, S. G., Lu, X.-W., and Horvitz, H. R. (1994). *Genetics* **137**, 987.
18. Granato, M., Schnabel, H., and Schnabel, R. (1994). *Development* **120**, 3005.
19. Miller, D. M. Shen, M. M., Shamu, C. E., Burglin, T. R., Ruvkun, G., Dubois, M. L., Ghee, M., and Wilson, L. (1992). *Nature* **355**, 841.
20. Bloom, L. and Horvitz, H. R. (1997). *Proc. Natl Acad. Sci. USA* **94**, 3414.
21. Jin, Y. S., Hoskins, R., and Horvitz, H. R. (1994). *Nature* **372**, 780.
22. Stone, S. and Shaw, J. E. (1993). *Gene* **131**, 167.
23. Maduro, M. and Pilgrim, D. (1995). *Genetics* **141**, 977.
24. Yochem, J., Gu, T., and Han, M. (1998). *Genetics* **149**, 1323.
25. Naito, M., Kohara, Y., and Kurosawa, Y. (1992). *Nucleic Acids Res.* **20**, 2967.
26. Maricq, A. V., Peckol, E., Driscoll, M., and Bargmann, C. I. (1995). *Nature* **378**, 78.
27. Williams, W. D. (1995). In *Methods in cell biology:* Caenorhabditis elegans: *modern biological analysis of an organism* (ed. H. F. Epstein and D. C. Shakes), p. 81. Academic Press, San Diego, CA.
28. Barnes, T. M., Jin, Y., Horvitz, H. R., Ruvkun, G., and Hekimi, S. (1996). *J. Neurochem.* **67**, 46.
29. Kelly, W. G., Xu, S., Montgomery, M. K., and Fire, A. (1997). *Genetics* **146**, 227.

6

Reverse genetics

1. Introduction

The nematode *Caenorhabditis elegans* has about 19 000 genes. Approximately 2000 of these have been identified in forward genetic screens for animals with mutant phenotypes. The generation and analysis of such mutants has been the key to the success of the *C. elegans* model because the *in-vivo* function of a gene can be inferred from its mutant phenotype. In principle, it is possible to identify mutations in the remaining 17 000 genes using various forward genetic screens. In practice, however, it is unlikely that forward genetics will lead to such an end in the foreseeable future. Further, now that nearly all the sequence of *C. elegans* is known, many genetics projects run in reverse: where a gene is identified first through its sequence and then mutations are recovered that inactivate or alter its function.

Several imperatives have driven the effort to develop strategies for the targeted inactivation of genes in *C. elegans*. One might expect that such strategies would include recombination-mediated gene replacement, where a wild-type copy of a gene is replaced with a mutant copy. Regrettably, recombination-mediated gene replacement, though effective in many other systems, has not worked with *C. elegans*. However, this chapter will describe four strategies that have worked. First, I will describe a method for the epigenetic inactivation of genes using double-stranded RNA from the target locus. Following this I will describe three strategies to generate germline mutations at specific target loci. Each of the latter involves a mutagenesis step followed by the selective identification of animals carrying mutations at the target locus.

2. Epigenetic gene inactivation

2.1 RNA-mediated gene inactivation (RNAi)

In a study of the *C. elegans* gene *par-1*, Kenneth Kemphues and co-workers noticed that either the sense or antisense transcript from *par-1*, when injected into the germline, led to the epigenetic inactivation of the gene in the resulting progeny (1). Subsequently, this phenomenon was extended to other genes. Andrew Fire, Craig Mello, and their colleagues showed that the phenomenon is strongest with double-stranded RNA (2). The mechanism, though not fully understood, may be distinct from the supposed mechanisms for the inactivation of genes via antisense RNA, and therefore the strategy has been given a distinct name, RNA-mediated interference (RNAi) (2, 3). RNA-mediated interference, using double-stranded (ds) RNA from particular target loci, is now routinely performed in *C. elegans* to determine the likely loss of function phenotype of a gene or to test whether the phenotype of a particular genetic mutant corresponds to the RNAi phenotype of a cloned gene. As this technology is in its early stages, improvements to the protocol are inevitable. Therefore, I will give only a general overview of the strategy and refer the reader to the following electronic address for further current details: `http://www.dartmouth.edu/artsci/bio/ambros/protocols/worm_protocols.html` (Andrew Fire, personal communication).

To induce RNA interference, dsRNA is injected, using a microcapillary pipette, into the gonad arms of young adult *C. elegans*. The injection protocols are similar to that used for DNA transformation (see Chapter 5). The dsRNA is made by synthesizing both the sense and antisense strands from a cDNA or cloned gene using bacteriophage RNA polymerases. The homologous single-stranded RNAs are annealed to create dsRNA. In initial experiments to determine whether RNAi induces a phenotype, it is not necessary to remove the template DNA from the injected RNA. In rare cases when the DNA template is not removed before injection, however, a heritable phenotype is transmitted through several generations, presumably due to the opportunistic transcription of sense and antisense RNA from the inherited template DNA. Though rare, the heritable phenotypes may not be the same as the acute phenotypes and may not correspond to the null phenotype. Therefore, the template DNA should be removed from the injected RNA before detailed phenotyping.

Many genes are susceptible to RNAi including genes expressed in the germline, the early embryo, and throughout larval development. The interference may be seen both acutely in the tissues of the injected animal, and in their progeny (4). To monitor phenotypes in the progeny, the injected animals are transferred to fresh culture plates where they are allowed to lay eggs for 24 hours. They are then transferred again for a second 24-hour brood. Surprisingly, it is not essential to inject the RNA into the gonad to induce interference; injections anywhere in the body cavity can induce an RNAi phenotype in the progeny (4). This shows that the interfering RNA can cross cell boundaries.

In some cases up to 100% of the offspring from an injected animal shows RNAi-induced phenotypes. In such cases the RNAi methods are robust, straightforward, and reliable. More challenging, however, is phenotype analysis. Guidelines for phenotype analysis are given below in Section 4.

3. Mutational germline gene inactivation

RNAi works for many but not all genes, and it does not always reveal the null phenotype. Furthermore, it does not provide the full range of genetic options that up to now have made *C. elegans* such a successful genetic model system. These include the ability to undertake genetic screens to identify suppresser or enhancer loci, and the generation of genetic mosaics to study the tissue-specific functions of essential genes. The following three strategies, see section 3.1, 3.2.1 and 3.2.2, allow for the mutational inactivation of genes in the germline of *C. elegans*.

3.1 Phenotype-dependent mutational screens

In principle, if one could foresee the phenotype of an animal with a mutation at a particular locus, then one could recover such mutants by screening for that phenotype. The types of data that might influence one's guesses as to the mutant phenotype include the expression pattern of the gene or whether the gene is known to act in an already well-characterized pathway. Clearly, however, such a prediction inspires little confidence in the absence of other data. Using RNAi, however, one can often discover the likely loss of the function phenotype of a particular locus. This information can form the basis for standard mutant hunts.

To illustrate this strategy I will described the recovery of mutations in the *C. elegans* gene, *deb-1*, which encodes a homologue to the vertebrate focal-adhesion protein, vinculin (4, 5). *C. elegans* vinculin is found in structures in the body-wall muscle known as dense bodies. Dense bodies function to link actin filaments in the muscle to the adjacent plasma membrane. The *deb-1* gene was cloned using antibodies to screen a cDNA expression library. After finding its location on the *C. elegans* physical map and correlating this physical interval with an interval on the genetic map, it was discovered that there were no known mutations in the *deb-1* gene. In a determined effort to apply genetic methods to the study of vinculin, a genetic screen was devised to recover *deb-1* mutants (6). It was predicted that mutations in the *deb-1* gene would be homozygous lethal, with a distinctive arrest phenotype like that of another essential muscle-affecting gene, *myo-3*, which encodes a myosin heavy chain (7).

This prediction formed the basis for a multilevel genetic screen. First, animals with mutations in essential genes within a several map-unit interval surrounding the location of the *deb-1* gene were isolated (see *Figure 1*). To do so, a strain was constructed that carried the genetic markers *unc-5* and *unc-44*, which flank the *deb-1* gene. Mutations in these *unc* marker genes cause animals to move abnormally. Homozygous *unc-5* animals can be readily distinguished from homozygous *unc-44* animals. P_0 animals were treated with the mutagen ethylmethanesulfonate (EMS) according to standard procedures (see Chapter 12). Approximately 2000, phenotypically wild-type, F_1 self progeny were picked on to 2000 fresh culture plates. After the F_1 had reproduced by self-fertilization, F_2 progeny were scored for the presence or absence of the Unc animals. Most of the F_1 clones segregated three classes of progeny: wild-type, Unc-44, and Unc-5 (see *Figure 1A*). These were discarded. In several cases, however, one of the Unc classes was missing (see *Figures 1B* and *C*). This was diagnostic for the presence of an induced lethal mutation linked to the missing Unc class. Mutants from this lethal class were examined to identify those that had arrest phenotypes that matched *myo-3* mutants. Finally, using antibodies to the *deb-1* product, vinculin, mutants were identified, within this restricted set, that failed to make the protein (see *Figure 1C*). These mutants were later confirmed by transformation rescue to be a result of mutations in the *deb-1* gene.

3.2 Phenotype-independent mutational screens

The strategy described in Section 3.1 works if appropriate sets of linked markers exist that will not interfere with the identification of the target phenotype, and when the target phenotype can be scored readily. Often this is not the case. Many genes in *C. elegans* have functions that wholly or partially overlap the function of other genes. The elimination of such genes by mutation often leads to animals with phenotypes that cannot be readily distinguished from wild-type using standard assays under standard laboratory culture conditions. Superficially, one might question the value of such a mutant that lacks an easily scored, informative phenotype. The answer is context-dependent. In some cases one may have reasons to suspect redundancy, e.g. when other known genes are recognized to have significant homology to the target. In such cases eliminating the function of all genes in the family might be necessary to reveal a phenotype. In other cases one may not have a ready rationale for the absence of a phenotype of a particular null mutant. Nevertheless, such mutants are still valuable, i.e. they provide a strain that can be used in standard genetic screens to recover mutations at loci that produce a phenotype only in the absence of the first gene, so-called synthetic phenotypes.

The methods described below allow for the functional knockout of any gene, including those that are members of gene families in which the phenotypes of any single gene knockout cannot be scored readily. Both strategies make use of PCR to detect rearrangements at specific target loci. A PCR phenotype is dominant; thus one can detect mutations in animals that are heterozygous for the rearrangement. Therefore, one can capture viable animals carrying only a single copy of a mutation in an essential gene that is lethal when homozygous.

3.2.1 PCR identification of transposable-element induced rearrangements (PITR)

C. elegans contains several mobile DNA elements that excise and insert from

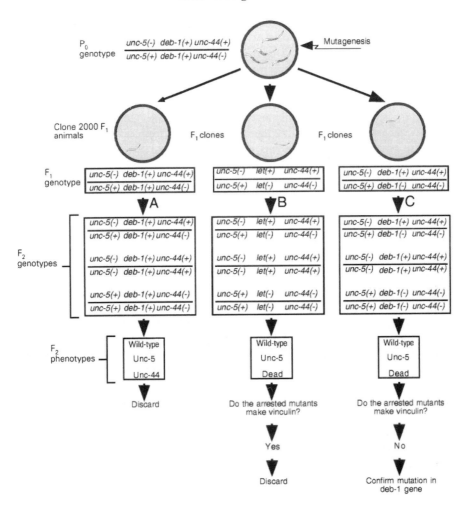

Figure 1. Screens for mutations with a predicted phenotype. The successful identification of mutations in the *deb-1* gene was a result of the successful prediction of the phenotype of a *deb-1* mutant. First, with the expectation that *deb-1* mutants would be homozygous lethal, screens were structured to identify mutations in any lethal gene in the region around *deb-1*. This was done by constructing and mutagenizing a strain that was doubly heterozygous for mutations in the genes *unc-5* and *unc-44*, cloning the F_1 descendants after the mutagenesis, and scoring the progeny that segregated from the F_1 clones. Most clones segregated progeny with all of the genotypes represented in the parents (path A). Those exceptional clones that failed to segregate one or the other of the expected Unc classes of marker mutants (paths B and C) were studied further. In a few cases the absence of an Unc class was due to recombination in the parental germline. In many other cases, however, the absence of the Unc class was the result of the induction of a new mutation in an essential gene in the region of *deb-1*. One of the new lethal mutants identified in this way was subsequently shown to lack the *deb-1* gene product, vinculin, and to have mutations in the gene (path C).

101

and to various sites throughout the genome. The 1.6-kb transposable element *Tc1* is the most thoroughly studied of these. Some strains of *C. elegans* are relatively more active for *Tc1* transposition. These so-called mutator strains show a high rate of spontaneous mutation. Rushforth *et al.* demonstrated that one could recover animals that carry *Tc1* at specific target loci from populations of animals where the *Tc1* is actively transposing (8). The strategy for identifying such mutants is displayed in *Figure 2A*. As the details are similar to the methods described in more detail in Section 3.2.2, a complete description will not be given in this section. There are two other reviews that describe the process in detail (9, 10). Briefly, small populations of animals from a mutator strain are grown for several generations. Portions of these populations are harvested and the DNA from these animals is prepared for PCR. To identify those populations that contain animals with *Tc1* at a particular target locus, one performs PCR with oligonucleotide primers that allow for the amplification of the DNA extending from the *Tc1* to the surrounding target gene. When a *Tc1* is identified in a particular DNA preparation, siblings of the animals that gave rise to that DNA are recovered from the original population. This population is then divided into smaller populations and the cycle is repeated. In this way one can eventually identify a clone that carries a *Tc1* in the gene of interest.

One might reasonably expect that the insertion of a 1.6-kb transposable element into the coding region of a gene should inactivate that gene. To the surprise of all, however, in those cases when such *Tc1* insertions are identified by PCR, typically, the insertions do not lead to gene inactivation. Rushforth and Anderson studied one such case, and found that the *Tc1* was post-transcriptionally spliced out of the message (11). After much enthusiasm, therefore, this strategy lost favour. The enthusiasm was re-ignited, however, when Zwaal, Plasterk and their colleagues discovered that, even in cases where the insertion of *Tc1* did not inactivate the disrupted gene, the resulting strain could be used to generate a deletion of the gene (12). Thus, one begins with such an insertion strain, and looks to identify animals in which the *Tc1* has excised imprecisely and taken part of the gene with it. As with screens to identify a *Tc1* insertion strain, PCR is used to identify such a deletion strain (see *Figure 2B*). Often this second step is a challenge, however, because in the process of isolating a strain with a *Tc1* insertion, one often unintentionally selects for healthy animals that have lost transposon activity. Furthermore, as this method generally takes two steps, many investigators use the method described in Section 3.2.2, and search directly for deletions caused by chemical mutagens.

3.2.2 PCR identification of chemically induced rearrangements (PICR)

PCR can be used to identify animals with deletions at a specific locus in populations of animals treated with chemical mutagens. *Protocols 2* and *3* describe

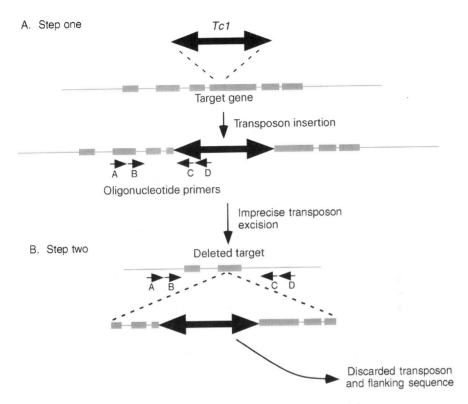

A. Step one

Tc1

Target gene

Transposon insertion

A B   C D

Oligonucleotide primers

Imprecise transposon
excision

B. Step two

Deleted target

A B   C D

Discarded transposon
and flanking sequence

Figure 2. Using PCR to identify a transposon insertion at a target gene. Typically, eliminating the function of a gene using the transposon method requires two steps. First, nested primers that fall within the transposon *Tc1* and the gene target are used in PCR screens of many small populations of animals that show a high rate of *Tc1* transposition. Second, once a strain is isolated with a transposon in the target gene, PCR screens are performed to identify populations with animals where the transposon has excised, taking part of the gene with it. A PCR fragment smaller than the wild-type fragment is diagnostic for such a mutant.

the use of ethylmethanesulfonate (EMS), diepoxybutane (DEB), and UV/trimethylpsoralen (UV/TMP) to induce deletions that can be detected by PCR. These three mutagens have previously been well characterized with regard to their acute toxicity to the mutagenized animals and to the frequency of mutagenesis in the progeny (1). It is clear, however, that the recovery frequency for PCR detectable deletions is much lower than the overall reported mutation rate with these compounds. It appears that only a fraction of the mutations caused by these chemicals lead to a rearrangement detectable by PCR with template DNA from large mixed populations. In our experience we obtain the highest frequencies with UV/TMP and DEB, and relatively lower frequencies with EMS.

The use of chemicals to induce mutations in *C. elegans* requires that one controls the level and length of time of exposure to the mutagen as well as the stage at which the animals are exposed to the agent (13) (see below). Ideally, one should gauge the level of exposure to the mutagen by balancing the forward rate of mutagenesis at the target locus against the rate at other unrelated loci. However, since one cannot directly measure the background mutational load generated by a particular mutagen, and since establishing a cut-off for an acceptable background rate is arbitrary, the level of mutagenesis is generally based on practical experience.

As chemicals do not cause deletions directly, but rather cause lesions in DNA that disrupt the DNA repair or replication machinery thereby leading to a deletion, the most appropriate stage for exposure to the mutagen is during the period where the mutagenesis target, the DNA in the germline, undergoes replication. In *C. elegans* this is during the third and fourth larval stage (L3/L4) when the germline undergoes mitotic expansion prior to meiosis. Large numbers of L3/L4-staged animals can be isolated by treating an asynchronous mixed population of animals with basic hypochlorite (see *Figure 3*) (see *Protocol 1*).

Hypochlorite kills all adult- and larval-staged animals in the population, but spares the developing embryos that are encased in a relatively impervious eggshell. The eggs are cultured for 52 h, during which time they hatch and develop relatively synchronously to the third or fourth larval stage (L3/L4). The staged larvae are then treated with chemical mutagens (see *Protocols 2* and *3*). Often, in the construction of a mutant bank, we split the harvest and treat with two mutagens. Following mutagenesis, the animals are cultured for an additional 24 h, after which they become adults with eggs. These eggs, which contain the F_1 progeny of the mutagenized animals, are harvested (see *Protocol 4*) following a second treatment with basic hypochlorite (see *Figure 3*). The F_1 larvae from the mutagenesis are cultured in many small, discrete populations (see *Protocol 5*). Though there are no hard-and-fast rules for the numbers of such populations that one should seek to produce for any single mutant library, we routinely start 1152 populations with 500 F_1 animals per culture for a total of 575 000 F_1 animals. *C. elegans* is diploid and so this represents 1 152 000 mutagenized genomes.

At present, using current protocols, the odds of recovering a mutation in any given target from such a library are about 50:50. Nevertheless, the work required to establish such a mutant bank is well within the range of most small laboratories. Further, it is reasonable to make another such mutant library should one fail to identify the desired mutants in the first instance. Substantially more effort is required in the subsequent sib selection steps. Investing relatively more effort in setting up the mutant bank, so that the size of the starting populations is greatly reduced, could lower the effort required for the sib selections. If one wishes to capture deletions in only a few genes, however, the guidelines below strike a reasonable balance of effort at each stage.

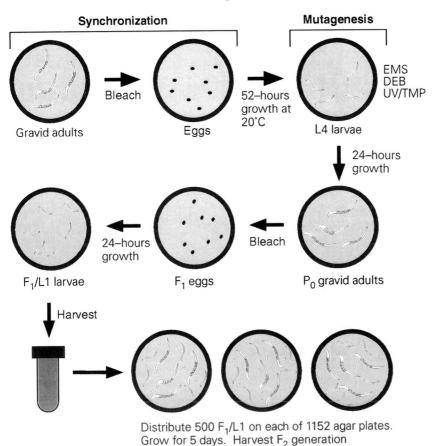

Distribute 500 F_1/L1 on each of 1152 agar plates.
Grow for 5 days. Harvest F_2 generation

Figure 3. Setting up and executing a chemical mutagenesis. Animals are synchronized for mutagenesis by treating a population of gravid hermaphrodites with a hypochlorite solution (bleach). The surviving eggs are cultured for 52 hours, during which time most of the embryos develop, hatch, and grow to the L4 stage. A portion of the synchronized culture is treated with either EMS, DEB, or UV/TMP, chemical mutagens that are known to cause deletions. After an additional 24-h culture, the F_1 descendants from the mutagenized animals are harvested with a second hypochlorite treatment. The F_1 eggs are cultured for 24 hours after which they develop and hatch. The resulting F_1 larvae are harvested, counted, and distributed on 1152 fresh culture plates, with 500 animals per plate.

The initial F_1 pools are cultured for 5 days, after which most of the animals in each population represent their F_2 descendants. A portion of each population is harvested (see *Protocol 6*), and DNA is then prepared for PCR (see *Figure 4*). The culture plate is flooded with sterile water, and approximately 25% of the worms are transferred in water to a fresh microtitre well.

The 75% of animals that remain on the culture plate are reserved for the

105

Grow worms on 1152 60-mm RNGM–agarose plates

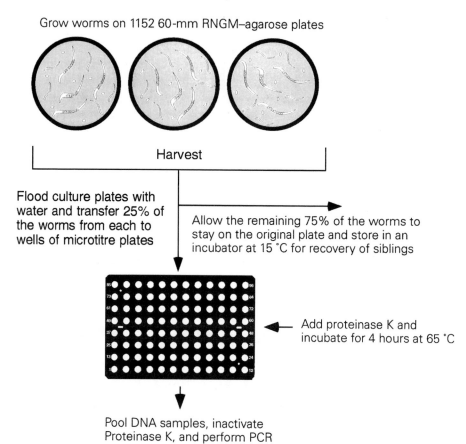

Figure 4. Harvest and template preparation. The culture plates are flooded with sterile water and portions of the population are transferred to the wells of a 96-well microtitre plate. A solution containing detergent and Proteinase K is added to the worms. The worms are frozen, thawed, and incubated at 65°C for 4 hours. After pooling and heating to inactivate the Proteinase K, the samples are ready for PCR.

next steps, i.e. where siblings of the animals in the DNA preparation are harvested to capture the candidate mutant. In principle, these populations could be equilibrated with a freezing solution and the animals stored long-term at –80°C. Often, however, only a small percentage of animals from a given population survive freezing, and therefore when starting with a complex population it is unlikely that one can reproducibly recover animals representing all genotypes in the population. When simply left on the culture plates, the sibling animals enter a long-lived developmental arrest stage called dauer. Dauer larvae can be stored for 4–6 weeks at 15°C with little significant loss in viability. One can screen for deletions at many loci within 6 weeks, and so for small numbers of genes freezing the sibling populations is neither productive nor desirable.

When the harvest is complete, portions of each of the original 1152 worm cultures are represented in 12 microtitre plates. The DNA from the harvested worms is prepared for PCR using a modification of the procedure described by Barstead and Waterston (8) (see *Protocol 7*). An equal volume of a solution containing Proteinase K and non-ionic detergents is added to the worm suspension. The filled microtitre plates are incubated in a 65°C oven for 4 h. After heating to 95°C to inactivate the Proteinase K, the DNA is ready for the PCR reactions.

To detect deletions at a target locus, two rounds of PCR are performed with a nested set of primers that produce a wild-type PCR fragment of about 2–3.5 kb (see *Figure 5*). Putative deletion fragments are identified by the

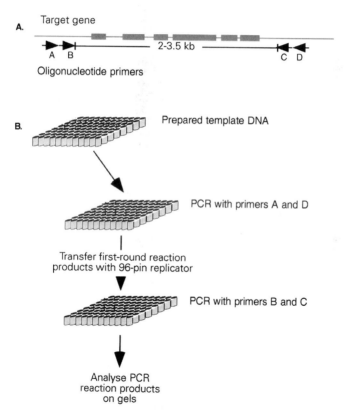

Figure 5. Primer selection and PCR reactions. A nested set of PCR primers that flank the target are used in two reaction sets. The inside primer sets are selected to be 2–3.5 kb apart. Prepared template DNA is mixed with the first-round reaction components, and incubated in a thermocycler for 35 cycles. The first-round reaction products are transferred to the second-round reaction components using a pin replicator. Only the second-round reaction products are analysed on 1% agarose gels.

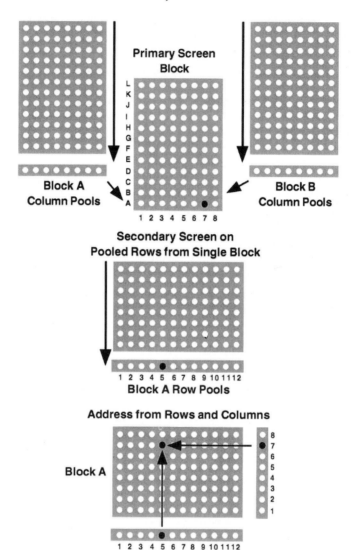

Figure 6. Pooling the DNA samples. First, each of the 12 blocks is reduced to a single set of 8 samples by pooling the columns. All 12 microtitre blocks, then, are represented in a single set of 96 reactions. The initial screens are performed on the pooled columns. The identification of a deletion fragment in the initial PCR reactions, therefore, leads to a row of 12 samples in a single block. The pooled rows from only the candidate block are tested using the same PCR primers to identify the address of the candidate population. The appearance of the deletion fragment in both the pooled columns and rows is taken to confirm the legitimacy of the candidate, and rule out false-positives. It is essential that the fragment sizes of the candidate sample from the rows and columns are directly compared on the same gel.

presence of a PCR fragment that is shorter than the wild-type fragment. Often, one detects false-positive fragments, but these are readily identified as such because a false deletion fragment is not repeatedly observed in follow-up PCR reactions. Fragment sizes are assessed electrophoretically on standard 1% agarose gels that are configured to accept samples in sets of 12 from a standard 96-well microtitre plate. Rather than testing all 1152 samples individually, the samples are pooled strategically (see *Protocol 8*) in a way that balances the need for a highly sensitive and specific assay against the desire to identify the particular source of the positive sample with a minimal number of PCR reactions (see *Figure 6*) (see *Protocol 9*). We have shown that one can reproducibly identify a particular deletion chromosome, even when that chromosome is in the presence of up to 20000 chromosomes that are wild-type at the target locus. To determine the specific address of a particular sample in the microtitre array, the columns from the initial 12-plate sets are pooled into a single microtitre plate for the initial PCR screens (see *Protocol 10*). The observation of a smaller than normal PCR fragment is taken as presumptive evidence that the source population carried animals with a deletion at the target locus. The exact address of the candidate population is confirmed with additional PCR reactions on the 12 DNA samples that represent the pooled rows from the original 12-plate set (see *Figure 6*).

The reserved sibling population that corresponds to the PCR address is harvested, divided into smaller sub populations, allowed to reproduce, and the process is repeated to capture the animals that carry the deletion. Typically, the complexity of the pools starts at 500 animals in the initial round of screening, and is then reduced to 50, 10, and a single animal in the first (see *Protocol 11*), second (see *Protocol 12*), and third (see *Protocol 13*) rounds of sib selection, respectively.

Updates of the following protocols are maintained at: `http://snmc01.omrf.ouhsc.edu/revgen/RevGen.html`

Protocol 1. Growth synchronization *(see Figure 3)*

Equipment and reagents

- Rich NGM (RNGM) plates: 3 g NaCl, 7.5 g Bacto-peptone[a] and 15 g agarose per litre. Autoclaved and cool to 55°C. Then add 1 ml cholesterol (5 mg/ml in ethanol, flammable!), 1 ml 1 M CaCl$_2$, 1 ml 1 M MgSO$_4$, 25 ml 1 M KH$_2$PO$_4$, pH 6.0.
- *E. coli* OP50
- M9 buffer: 22 mM KH$_2$PO$_4$, 22 mM Na$_2$HPO$_4$, 85 mM NaCl, 1 mM MgSO$_4$
- Basic hypochlorite solution: 0.25 M KOH, 1–1.5% hypochlorite, freshly mixed
- 15 cm Petri plates

Method

1. Grow worms on 5–10 15-cm RNGM plates to generate a large population of gravid adults. Depending on the size of the starting population this should take from 3 to 5 days.

Protocol 1. *Continued*

2. Wash the gravid adults off the plates and treat with basic hypochlorite to harvest the eggs. Collect worms by centrifugation in M9 Buffer, add 10 volumes of basic hypochlorite, and incubate at room temperature for about 4 minutes. Collect eggs (and some residue of carcass) by centrifugation (400 *g*, 5 min, 4°C).

3. Wash the eggs 5 times with 10 vol. of M9 buffer and collect by gentle centrifugation.

4. Distribute the eggs across 10–20 15-cm plates containing RNGM medium seeded with an *E. coli* OP50 lawn (see Chapter 4, Section 3.3).

5. Culture the worms for 52 hours at 20°C.

[a] The peptone is 3 × standard NGM.
[b] Worms grown on plates made with agar often burrow under the surface. Burrowing causes problems with the harvest. Using agarose instead of agar prevents burrowing. Not all brands of agarose work for this purpose. We recommend Electrophoresis Grade Agarose from FisherBiotech (cat. no. BP160–500).

Protocol 2. UV/TMP mutagenesis (14)

Equipment and reagents

- M9 buffer (see *Protocol 1*)
- 3 mg/ml 4,5′,8-trimethylpsoralen (TMP) (Sigma, cat. no. 6137) in DMSO
- Two hand-held UV lamps (Fisher Scientific, cat. no. 11–984–40, Model UVL-21 Blak-Ray Lamp, long-wave UV—365 nm) mounted using appropriate clamps on a ring stand

- UV dose meter (Fisher Scientific, cat. no. 97–0015–01, Model UVX Digital Radiometer with UVX-36)
- UV safety equipment
- 50 ml centrifuge tubes, capped
- Sterile, glass Petri plates

Method

1. Collect the synchronized culture of worms (see *Protocol 1*) by washing the culture plates with M9 buffer.

2. To an 8 ml suspension of worms in M9 buffer, add 8 ml of M9 buffer containing TMP at 60 µg/ml.

3. Incubate the worm suspension in a horizontal, capped 50 ml centrifuge tube at room temperature in the dark for 15 min.

4. Transfer the worms in the dark to a sterile, glass, 15 cm Petri plate.

5. Irradiate the suspension with 360 nm UV light for 90 sec at 340 µW/cm with gentle shaking. Adjust the height of the lamps to generate the appropriate UV dose as measured with a UV dose meter.

Protocol 3. EMS mutagenesis/DEB mutagenesis

Equipment and reagents

- M9 buffer (see *Protocol 1*)
- EMS (methanesulfonic acid ethyl ester) (Sigma, cat. no. M-0880)

Or

- DEB (1,2:3,4-diepoxybutane) (Sigma, cat. no. D7019)
- 50 ml centrifuge tubes, capped

Method

1. To an 8 ml suspension of worms in M9 buffer, add 8 ml of M9 buffer containing 100 mM EMS *or* 0.2 mM DEB.

2. Incubate the worm suspension in a horizontal, capped 50 ml centrifuge tube at room temperature for 4 h (for EMS) or 3 h (for DEB) with moderate shaking.

3. Wash the worms five times with 10 vol. of M9 buffer. After each wash collect the worms by centrifugation at 1000 g at 4°C for 5 min.

Protocol 4. Collection of F_1 larvae (see *Figure 3*)

Equipment and reagents

- RNGM plates seeded with *E. coli* OP50 and unseeded
- M9 buffer and alkaline hypochlorite (see *Protocol 1*)

Method

1. Plate the mutagenized worms (from *Protocol 2* or *3*) on 15–20 seeded 15 cm RNGM plates.

2. Culture the worms for 24 h. (Keep TMP-treated worms in the dark.)

3. Collect F_1 eggs by treating the worms with basic hypochlorite as described in *Protocol 1*.

4. Allow the eggs to hatch on the surface of an unseeded 15 cm RNGM plate.

Protocol 5. Plating the mutant library

Equipment and reagents

- M9 buffer (see *Protocol 1*)
- 6 cm, *E. coli* OP50-seeded RNGM plates (see *Protocol 1*)
- A repeating pipette with sterilized syringes
- Lidded plastic containers

Protocol 5. *Continued*

Method

1. Collect the F_1/L1 worms (from *Protocol 4*) in M9 buffer.
2. Determine the concentration of worms in suspension by counting the number of worms in a measured aliquot. Serial dilutions may be necessary to make the counts reasonable and accurate.
3. Adjust the concentration of the suspension to 500 worms/0.1 ml.
4. Distribute the worms in groups of 500 on to 1152 seeded RNGM plates using a repeating pipette.
5. Store the plates in groups of 96 in lidded plastic containers.
6. Culture the worms for 5 days at 20°C.

Protocol 6. Library harvest

Equipment and reagents

- Repeating pipette
- Streptomycin
- Mycostatin
- Deep, 96-well, microtitre plates

Method

1. Number the plates from *Protocol 5* in groups of 96 and place them in stacks of six.
2. Use a repeating pipette to add 0.75 ml of sterile distilled water containing streptomycin (100 μg/ml) and mycostatin (12.5 μg/ml) to each plate in a single stack of six.
3. Rock the plates to dislodge the worms and, using a standard micropipette, transfer 150 μl of the worm suspension from each of the six plates to the appropriate six wells of a deep, 96-well, microtitre plate (see *Figure 4*). Keep track of the well position by temporarily ejecting the spent micropipette tip into the microtitre well.

Protocol 7. DNA preparation

Equipment and reagents

- Proteinase K solution: 50 mM KCl, 10 mM Tris–HCl pH 8.3, 2.5 mM $MgCl_2$, 0.45% NP-40, 0.45% Tween-20, 0.01% gelatin, 200 μg/ml Proteinase K
- Flexible mat lid, 96-well (Midwest Scientific, cat. no. P9618)

Method

1. After filling each 96-well microtitre block with worms (see *Protocol 6*), add 150 μl of the Proteinase K solution to each well.

2. Seal the blocks with a mat lid.

3. Freeze the worms at $-80\,°C$ for 20–30 min.

4. Incubate the blocks, with intermittent mixing, for 4 hours at $65\,°C$. Seal the perimeter with tape prior to a prolonged incubation if the lids do not completely seal the wells. Avoid vigorous mixing that may lead to well-to-well contamination. Make absolutely sure that the DNA is completely dispersed in the solution; this is critical.

Protocol 8. Sample pooling

Equipment and reagents

- PCR tubes in strips of 12 (FisherBrand 12-Strip, 0.2 ml, thin-wall reaction tubes (cat. no. 05–407–3A), with caps (cat. no. 05–407–4B))
- A multichannel pipette
- PCR tubes in strips of eight tubes (MJ Research 8-Strip, 0.2 ml reaction tubes (cat. no. TBS-022), with caps (cat. no. TCS-0801))
- Deep, 96-well microtitre plates

Method

1. Dilute each of the 1152 DNA samples from *Protocol 7* with an equal volume of sterile distilled water.

2. Pool both the rows and columns from each of the 12 plates, such that the 12 initial plates are reduced to two sets of 96 samples (see *Figure 6*). Use PCR tubes in strips of 12 for the pooled columns. Use strips of eight tubes for the pooled rows.

3. Use a multichannel pipette to transfer 50 µl of a given sample (in sets of 8 or 12) to the appropriate wells of a deep 96-well microtitre plate.

4. Centrifuge the pooled samples to pellet the cuticle residue that remains after the Proteinase K treatment.

5. Transfer approximately 200 µl of each cleared sample to PCR tubes.

6. Heat the samples to $95\,°C$ for 20 min to inactivate the Proteinase K.

Protocol 9. PCR reactions

Equipment and reagents

- 96-well PCR trays (MJ Research, cat. no. MPL-9601)
- Dimpled rubber mats (Perkin Elmer 96-well plate cover, cat. no. N801–0550)
- *Taq* polymerase
- $10 \times$ dNTP stock: 2 mM each of dATP, dCTP, dGTP, dTTP
- $10 \times$ Assay buffer: 100 mM Tris–HCl pH 8.3, 500 mM KCl, 25 mM $MgCl_2$
- 96-pin replicator (Boekel Replicator; Fisher, cat. no. 05–450–9)
- Thermocycler
- Template DNA (from *Protocol 8*)

Protocol 9. *Continued*

Method

1. Select nested primer sets that are separated by approximately 2–3.5 kb (see *Figure 5*). Avoid primers that fall within a sequence which is repeated in the genome. Select primers that are about 20 bases long with about a 50% G/C content.

2. Perform the PCR reactions in 96-well PCR trays sealed with dimpled rubber mats.

3. For the first round PCR reaction combine: 5 μl template DNA, 10 pmoles of each outside primer, 2.5 μl of a 10 × dNTP stock, 2.5 μl of the 10 × Assay buffer, 0.5 units of *Taq* polymerase, add dH$_2$O to 25 μl.

4. Incubate the reactions in a thermocycler: 94°C for 30 sec 1 cycle followed by 94°C for 30 sec; 55°C for 20 sec; 72°C for 2 minutes, 35 cycles.

5. Using the pin replicator, transfer a small portion of the first-round PCR reaction to tubes containing 25 μl each of the second-round PCR reaction components. Adjust the reaction mix appropriately since the replicator transfers an insignificant volume to the reaction. If one insists on using a micropipette to transfer the first-round mix to the second-round reaction, dilute the first-round reaction 5–10-fold. Combine 10 pmoles of each inside primer, 2.5 μl of a 10 × dNTP stock, 2.5 μl of the 10 × Assay buffer, 0.5 units of *Taq* polymerase, dH$_2$O to 25 μl for each of the second-round PCRs.

6. Repeat the thermocycling described in step 4.

Protocol 10. Assaying the PCR reactions

Equipment and reagents
• An agarose gel system configured for the analysis of samples in microtitre arrays

Method

1. Perform the initial screens on 96 samples representing the pooled rows. Examine the PCR reaction products from *Protocol 9* on 1% agarose gels. Select samples which show bands that are smaller than the wild-type for further analysis. At this stage most of the small candidate PCR fragments represent false-positives.

2. Substantiate the validity of the deletion band and determine the precise address of the candidate sample by repeating the PCR on the 12 appropriate samples from the pooled columns. If a band appears in the appropriate samples after PCR of both the rows and columns, compare the sizes of the PCR fragments from the candidate samples

directly on the same gel. If the sizes are the same upon direct comparison, then the candidate is validated.

3. If needed, obtain further evidence of the validity of the candidate or produce template for sequencing by PCR from the core sample. The timely validation of the candidate and follow-through to the sib selection step is the key to success.

4. Proceed to the sib selections (see *Protocol 11*).

Protocol 11. First sib selection

Equipment and reagents

- M9 buffer (see *Protocol 1*)
- 12-well tissue culture plates (Evergreen Scientific, cat. no. 222–8046–01F) containing RNGM and seeded with *E. coli* OP50 (see *Protocol 1*)
- Equipment and reagents from *Protocol 6*
- Proteinase K solution (see *Protocol 7*)
- 12-channel pipette
- PCR equipment and reagents (see *Protocol 9*)

Method

1. Wash a portion of the worms off the original library culture plate with M9 buffer.

2. Count the worms to determine the concentration, dilute appropriately, and distribute the worms in sets of 50 into the wells of 16, 12-well, tissue culture plates. This gives a total of 192 populations.

3. Culture the worms at 20°C for 5 days.

4. Harvest a portion of each population for PCR. Flood the wells with 200 μl of sterile distilled water containing streptomycin and mycostatin as in *Protocol 6*. Transfer 50 μl of the worm suspension to a deep-well microtitre plate.[a]

5. Add 50 μl of Proteinase K solution, as in *Protocol 7*, to the 50 μl worm suspension and freeze the samples at −80°C for 20 min.

6. Incubate at 65°C for 4 h.

7. Add an equal volume of sterile distilled water to each sample and centrifuge to pellet the cuticle residue.

8. Transfer 5 μl of each sample to a 96-well PCR tray using a 12-channel pipette.

9. Heat the tray to 95°C for 15 min to inactivate the Proteinase K.

10. Add 20 μl of an appropriately prepared PCR reaction mix to each sample and repeat the nested PCR as in *Protocol 9*.

[a] It is convenient to use a repeating pipette to fill all 12 wells in a single plate with water, tip the plate to an angle of 45°, and extract the worm suspension using a standard micropipette. In this way, a practised technician can harvest 192 samples in 20 minutes.

Protocol 12. Second sib selection

Equipment and reagents

- 24-well tissue culture plates (Evergreen Scientific, #222–8044–01F) containing RNGM and seeded with *E. coli* OP50 (see *Protocol 1*)
- Deep, 96-well microtitre plates
- PCR equipment and reagents

Method

1. Harvest the worms from a candidate population, as in *Protocol 11*, and distribute in sets of 10 into the wells of four 24-well tissue culture plates. This gives a total of 96 populations.

2. Culture the worms for 5 days and harvest as in *Protocol 11*. Flood the wells with 100 μl of sterile distilled water and transfer 25 μl to a deep 96-well plate.

3. Process the DNA samples and perform PCR as described in *Protocol 11*.

Protocol 13. Third sib selection

Equipment and reagents

- 6 cm RNGM plates seeded with *E. coli* OP50 (see *Protocol 1*)
- PCR equipment and reagents
- 24-well tissue culture plates containing RNGM and seeded with *E. coli* OP50 (see *Protocol 1*)

Method

1. Harvest the worms from a candidate population and transfer a portion to several RNGM plates.

2. After 1 to 2 days pick single worms from these plates into the wells of four 24-well tissue culture plates for a total of 96 clones.

3. Harvest a portion of each population after 4–5 days and test with PCR as in *Protocol 11*.

---

4. Analysis of phenotypes

4.1 General considerations

The most challenging aspect of a reverse genetics project is phenotype analysis. In a standard mutant hunt one looks for animals with particular, narrowly defined phenotypes. Often, the ability to recognize such phenotypes is the result of an expertise gained from long experience with the model system. This expertise extends to a level of familiarity with the genetics and cell biology of the particular developmental pathway, uniquely allowing the investigator to make good scientific judgements, and to subsume and appropriately integrate new information. When one takes the reverse approach, however, it is possible that the investigator will be naïve with respect to the phenotype of the resulting mutant. Therefore, a reverse genetics project, should ideally lead to open collaborative research between groups that together have a wide range of appropriate expertise. However, some preliminary, phenotype analysis is necessary to make a judgement as to the most appropriate collaboration to pursue. This preliminary analysis may be greatly advanced if one knows which tissues express the target gene; one then can pay special attention to these tissues in the mutant. Before engaging in any phenotype analysis, however, one must eliminate background mutations. Crossing the mutant strain six to eight times to wild-type animals does this (see Chapter 12).

Even an inexperienced investigator can determine whether the mutation causes developmental arrest when homozygous. Developmental arrest is a phenotype that encompasses a broad range of gene products, and therefore it is difficult to generate an all-encompassing decision tree that can serve generally to direct one's analysis. However, it is relatively straightforward to determine the stage of developmental arrest and whether some of the more obvious morphogenic events take place. These events include the development of the gut and pharynx as well as the elongation of the embryo from an approximately spherical cluster of cells to a worm. Furthermore, by observing whether the animals are able to move, it can be determined whether the muscle or nervous system may be affected in the mutant.

Those mutations that are homozygous viable fall into two broad categories: morphological mutants and behavioural mutants. Recognizing the differences between the morphology or the behaviour of wild-type animals and that of a mutant is an iterative process. A feature that seems vaguely abnormal is often seen when examining a mutant. The wild-type animals must then be re-examined to determine whether the peculiarity truly represents a distinct phenotype or if it is present in wild-type animals but has not previously been understood.

Morphology can be examined through the relatively low resolution of the dissecting microscope to the highest resolution of the electron microscope (see Chapter 7). Phenotypical features seen under the dissecting microscope include body size and shape, and some changes in the positions or sizes of the internal organs. Defects in the hermaphrodite vulva and the male tail are often visible under the dissecting microscope. More subtle changes in morphology can be seen using the relatively greater resolution of differential-interference contract microscopy, including the presence of extra cells or cells that are misplaced. The use of fluorescence or other cytological tags, that can be assayed under light microscopy, permit the identification of those phenotypes that are based on changes in cell patterns. Mutant behaviours are

117

examined in live animals through the dissecting microscope (see Chapter 8). Such behaviours include relatively obvious changes in the rate or pattern of movement or, for example, subtle changes in the response to volatile odorants or touch.

These simple guidelines just touch the surface of a comprehensive and detailed phenotype analysis. To go deeper into the analysis, for example to determine whether a particular morphological defect is due to a change in the developmental lineage, the specification of cell fate, or the execution of a particular programme of differentiation, requires considerably more expertise. Moreover, a strategic plan should allow for the possibility that the phenotype of a given knockout worm may not be easily distinguished from wild-type. Since phenotype analysis can involve an immense variety and number of assays, it must be decided whether the analysis of a particularly recalcitrant mutant is sufficiently productive to warrant continuation. Finally, the analysis of partial loss-of-function or gain-of-function alleles often has a relatively greater impact on our understanding of the *in-vivo* function of the protein under study. Therefore, the analysis of the null phenotype should be considered to be only the beginning of a long-term, but inevitably fruitful, investment in this model system.

References

1. Guo, S. and Kemphues, K. J. (1995). *Cell* **81**, 611.
2. Fire, A., Xu, S., Montgomery, M. K., Kostas, S. A., Driver, S. E., and Mello, C. C. (1998). *Nature* **391**, 806.
3. Montgomery, M. K. and Fire, A. (1998). *Trends Genet.* **14**, 255.
4. Francis, G. R. and Waterston, R. H. (1985). *J. Cell Biol.* **101**, 1532.
5. Barstead, R. J. and Waterston, R. H. (1989). *J. Biol. Chem.* **264**, 10177.
6. Barstead, R. J. and Waterston, R. H. (1991). *J. Cell Biol.* **114**, 715.
7. Waterston, R. H. (1989). *EMBO J.* **8**, 3429.
8. Rushforth, A. M., Saari, B., and Anderson, P. (1993). *Mol. Cell. Biol.* **13**, 902.
9. Plasterk, R. H. (1992). *Bioessays* **14**, 629.
10. Plasterk, R. H. (1995). In *Caenorhabditis elegans*, Modern biological analysis of an organism, Vol. 48 (ed. Epstein, H. F. and Shakes, D. C.), p. 59. Academic Press, San Diego.
11. Rushforth, A. M. and Anderson, P. (1996). *Mol. Cell. Biol.* **16**, 422.
12. Zwaal, R. R., Broeks, A., van Meurs, J., Groenen, J. T., and Plasterk, R. H. (1993). *Proc. Natl Acad. Sci. USA* **90**, 7431.
13. Anderson, P. (1995). In *Caenorhabditis elegans*, Modern biological analysis of an organism, Vol. **48** (ed. Epstein, H. F. and Shakes, D. C.), p. 31. Academic Press, San Diego.
14. Yandell, M. D., Edgar, L. G., and Wood, W. B. (1994). *Proc. Natl Acad. Sci. USA* **91**, 1381.

7

Microscopy

RALF SCHNABEL

1. Introduction

This chapter is designed to help you to analyse the phenotype of your favourite worm or embryo by looking through a microscope. The worm is perfectly suited to analysis by light microscopy. The dimensions of the embryo (about 55 μm) or the adult (1 mm) are just in the right range, such that the whole embryo or significant parts of the body of larvae and adults can be viewed with the highest numerical aperture and therefore with the best achievable resolution. Thus, subcellular details can be seen without losing sight of the whole animal. Although a wealth of very helpful, molecular markers of cell fate are available (e.g. antibodies, transgenic lines with green-fluorescent protein (GFP)- or β-galactosidase-labelled proteins, or specific probes for RNA *in situ*), differentiated cells show very characteristic features when viewed directly through a light microscope fitted with Nomarski (DIC) optics, so that determining which tissue or organ a cell belongs to (e.g. pharynx, gut, hypodermis, nervous system, or body-wall muscle) is quite easy. The use of 4D-microscopy, which allows the development of whole embryos to be documented, adds a new level to the analysis of cells since the 'behaviour' of cells can also be followed. It is an interesting question whether the expression of a single molecular marker really defines the fate of a cell unambiguously. However, if a cell looks like, migrates, and intercalates like a hypodermal cell it is very likely a hypodermal cell.

Many investigators have now become interested in *C. elegans* because the function of their favourite molecule can be readily analysed in this nematode. It is therefore the general aim of this chapter to introduce molecular biologists to the basics of microscopy. To determine the function of the gene, at least its biological function as opposed to its 'chemical' function, one has to look through a microscope. Although this is mainly a book about methods, concepts are also very important methods and therefore a short conceptual discussion about what is required to determine the function of the gene, that is characterization of the 'phenotype' of the organism lacking a gene function, is included in the section discussing 4D-microscopy.

2. Stereomicroscopy

The handling of worms requires a stereo microscope. In my experience, the quality of the illumination in the base of the microscope is much more important than the actual optics of the microscope. You should use a sandblasted glass plate on top of the light source. When buying new microscopes it is advantageous to be able to compare the instruments side by side. You may be astonished how different the worms look under different makes of microscope. I prefer Wild microscopes, which now come under the name of Leica. The bases were called EB but are now called HL. It may be also worthwhile equipping an 'old, discarded' microscope with a new base. It should be possible to adapt new bases to old microscopes, but this may require some insistence. Good optics and a solid piece of fine mechanics never age. You may find that a good, old microscope has much better optics than a not so expensive new one.

With the breeding of transgenic worms expressing GFP fusion genes, it may be worthwhile to consider equipping a stereo microscope for fluorescence microscopy. It is now possible to use GFP constructs, for example, as transformation or genetic markers. You will find appropriate equipment on the market.

3. Light microscopy

3.1 Acquiring a new microscope

A higher power microscope is needed for close examination of worms. At this point you have to consider carefully how much you think you will need your microscope and what for. Exquisite optics mounted in a rather simple base yield the same pictures as a modern high-tech microscope, which needs several weeks' training to handle it properly. As a rule, it is much easier to simply turn a knob than to handle a motorized stage with a computer. It has almost been forgotten that modern films for classical photography still have a much higher resolution than even the most modern digital video camera. I give some photography tips later, but it may be worthwhile to consider using conventional cameras and digitizing images subsequently for processing and publishing. Although I build, as will be described later, the most complicated 4-dimensional microscopes I am still very much in favour of the classical mechanical microscope for everyday purposes. For the high-power microscope, spend your money on the optics not on the base.

Working with worms you will need a microscope equipped with differential interference contrast optics (DIC, Nomarski) and the attachments for fluorescence microscopy. The same lenses (Plan Neofluar) can be used for both purposes. There should be no phase rings in the lenses, as they will lower the resolution in either fluorescence or Nomarski microscopy. Usually you will only need a 5 × lens to find your specimen and to look at adults, as well as a 40 × oil (aperture 1.3) and a 100 × oil (aperture 1.3) to look at embryos, larvae, and adults. The rule for the highest magnification that can be achieved with a lens without loss of resolution, i.e. futile magnification, is aperture × 1000. Therefore, the 5 × lens covers the range from 50 to 120 x, the 40 × lens covers the whole range between 400 and 1000 x, and the 100 × lens the range above 1000 ×. With an additional 10 × lens (aperture 0.3) one can cover the whole range of magnification. To use the whole range of magnification an Optovar is required, which permits additional magnification either in discrete steps (1.25, 1. 6, and 2 times) or continuously with a zoom. I find a high-quality Optovar very useful since the worm specimen usually does not fit exactly to the standard magnifications of the microscope. The expense corresponds roughly to a good lens and you save more than this by using an Optovar. For capturing pictures with a video camera I use a combination of a 4 × expander (this depends on the size of the chip used in the camera) and an Optovar very successfully. When using a video camera it is crucial that your specimen covers a large part of the field of view seen by the camera, otherwise you do not use the resolution of the camera optimally (see below).

A condenser with an aperture of 0.9 is sufficient for the best Normarski. *C. elegans* adults and embryos are too thick for an oil condenser with an aperture of 1.4, and I have never seen any improvement. If you need to take Nomarski pictures of late larvae or adults you will need a condenser with a low aperture (0.3 or so) to be able to illuminate the whole field of view evenly. For good DIC/Nomarski optics it is essential to be able to turn the polarizing filter manually each time you set up your optics. Do not buy a microscope with fixed filters crossed at 90 degrees. Although theory requires this, in reality the best images are obtained with the filters turned slightly away from 90 degrees (see below). The microscope should contain a heat reflection filter to minimize damage to live specimens. A blue filter may also be useful to cut out long wavelength light. However, if you use a video camera check for the sensitivity of the camera over the spectrum, as some cameras are much less sensitive in short-wavelength light. In this case a blue filter is counterproductive because much more light is needed.

For fluorescence microscopy you need an appropriate lamp and filter sets. Go for 100-watt illumination. If you are concerned about too much light damaging your specimen there are now lamps whose intensity is continuously regulated. You need three filter sets: for FITC or GFP (blue, green), rhodamine (green, red), and DAPI (UV, blue). Although GFP can be observed using FITC filters, special filter sets for GFP are now available that reduce the background. I prefer filters with narrow transmission windows, although they are more expensive than those with wide wavelength ranges. Narrow filters will make the picture darker, but the signal-to-noise ratio appears better. In fluorescence microscopy the yield of light is obviously crucial. There are microscopes optimized for the light yield and others for resolution. I obviously opt for resolution. Modern films and video cameras are very sensitive. If your immunocytochemistry protocols are good you will very often have to lower the intensity of your light source anyway.

As with the stereo microscope it is a very good idea to compare different makes side by side with the same specimen. In direct comparison of different makes on several occasions, I have always reached the same conclusion: that Zeiss microscopes have the highest resolution. However, new microscopes are being developed all the time, and thus it is always worthwhile comparing the different makes that are available. Again insist, and never buy anything you can not inspect in your laboratory with your specimen. Also, although some representatives are much more active than others, there is generally no correlation to the quality of the microscopes. What you need is an excellent microscope in the end. A lot of information about microscopes and accessories can be found on the Web by searching with the term 'microscope'.

3.1.1 Comparing microscopes

(i) DIC (Nomarski) microscopes

Mount some embryos and larvae of all stages on an agar pad, as described below. It is also useful to cut some hermaphrodites and males into pieces and to include them in the preparation. The cuticle of adults is very rich in fine structures, which may be helpful in testing the quality of the optics. You should be able to see the alae and annuli of the L1 larvae at 1000 × magnification (see *Figure 1*). If you do not see them after setting the microscope up properly (see *Protocols 2 and 3*) look for another microscope. DIC microscopy is very sensitive to scattered light. The quality of the optics can be tested during the set-up procedure described in *Protocols 2 and 3*.

(ii) Fluorescence microscopes

Of course one will not buy two different microscopes for DIC and fluorescence microscopy. If you need Nomarski optics you should choose the microscope with the better Nomarski. Besides, in this way I think you will also get the microscope which has the best fluorescence optics. To test the fluorescence optics, early embryos stained for tubulin (monoclonal anti-α-tubulin, clone no. DM 1A; Sigma cat. no. T9026) will be very useful. The tubulin fibres are very good for seeing the resolution and the light sensitivity of the microscopes. If possible, take pictures for comparison. Tubulin fibres also make a good example for demonstrating that less light can give better resolution. The problem of exposure is discussed separately.

Again it is not necessary to have the latest, fancy computerized microscope. If you can acquire an 'old', well-equipped microscope, e.g. a Zeiss Photomicroscope III, take it. The numerical aperture is 1.4 in old and new microscopes. You may find that older microscopes have a nicer Nomarski and a better fluorescence than some modern instruments.

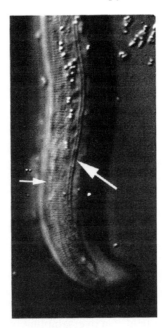

Figure 1. Testing Nomarski (DIC) optics. Surface of the head region of an L1 larva as seen with a 100 × Plan Neofluar (1.3) objective using a Zeiss Axioplan I microscope equipped with DIC. The picture was taken with a Hamamatsu Newvicon camera and digitized with a Perceptics PixelPipeline (PTP425) run in a Macintosh computer. Contrast and brightness were adapted with Adobe Photoshop. If the alae (large arrow) and the annuli (small arrow) cannot be seen in a microscope then the quality of the optics is insufficient. The picture also shows that a camera with a resolution of 700 lines is sufficient to depict the finest structures in a larva. One cannot see more looking directly through the microscope.

3.1.2 Which documentation system?

If you choose a modern, digital documentation system with a high-resolution digital camera, a motorized stage (or a functionally equivalent piezo mover) for z-series image acquisition, and deconvolution software—which, in turn, requires a very fast computer—the actual optics may be a minor part of the bill. Please consider very carefully what you will need and how often you will need it. I am firmly convinced that you do not need all this to capture the highest quality images. Digital recording *per se* using an analog or digital video camera is very useful if you need to analyse many different strains. Retrieving and looking at the pictures on the screen is very useful when comparing different experiments. Coupled to digital-picture processing and a rather costly video printer you can expect excellent results. Good advice and software concerning (low cost) digital microscopy can be obtained through the Research Services Branch (RSB) of the National Institute of Mental Health (NIMH), part of the National Institutes of Health (NIH) (Web site:

http://rsb.info.nih.gov/nih-image/about.html). The frame grabber digitizing the pictures is a very important part of the system. A very good camera in conjunction with a bad frame grabber gives very poor results. Consider very carefully whether it is worth obtaining a high-resolution digital camera. The resolution of a 100 × lens is realistically no better than 0.2 microns. If an embryo has a length of 60 µm with the field of view being collected having a length of 100 µm and this corresponds to approximately 750 pixels or lines of a camera, then you have an oversampling of approximately 1.5. In my experience this is sufficient. A higher resolution is not really useful and it takes a lot of effort to cope with the huge amount of data produced by the camera. It makes no sense to use only a small window on a high-resolution chip of a digital camera. If you make sure that the 768 pixels available with a 'normal' camera are used to picture a specimen which covers 60% of the optical field, then you have effectively the same resolution as with the best high-resolution camera. An important factor in the set-up is the size of the chip offering the resolution. As with direct observation, too high a magnification with an Optovar, in order to project your specimen to a reasonable size on the chip, just lowers the resolution again by blurring the picture. One has to make sure that the number of pixels corresponds to the resolution of the microscope and to the size of the observed field. An oversampling by a factor of three is more than sufficient. Also high-resolution cameras are still slow, and on many occasions I have found that the final printed images were not significantly better than those produced by good analog (tube or CCD) cameras in conjunction with a frame grabber (see also Section 7.2 for the problem of printing digital pictures). It only makes sense to use digital pictures if you are also prepared to purchase, or you at least have access to, a (very expensive) high-resolution video printer so that you can produce your pictures in hard copy. Bear in mind the necessity for a system to store and back up your data safely.

If you have only occasional need for a microscope then either consider whether somebody nearby has the appropriate equipment or stick to classical photography, which gives the very best resolution for a reasonable investment. Finally, some general advice about commercially available software packages that are often huge and very expensive. The range of possible image manipulations and fantastic overlays is very impressive. There are 3D-reconstruction and deconvolution programs to equip your normal microscope with all the features of confocal microscopy. In my experience, I have found no use for all these sophisticated features. I used the complicated software simply for saving the unmanipulated images to disk. One useful application, that produces beautiful pictures, is in colour coding and merging images collected from a double- or multiple-labelled specimen using different channels of the fluorescence microscope. This can be easily achieved using programs like Photoshop. But in the end you may publish the images in separate black and white glossy prints anyway, because of the very high charges most journals ask for colour figures.

In summary, your microscope set-up will only be as good as the weakest component in the system. The resolution of the camera has to be in scale with the optics. Do not trust calculations too much. In my experience, one has to experiment to see what gives the best results.

4. Nomarski microscopy

4.1 Mounting embryos or worms

The key to a good Nomarski image is a good agar pad upon which to mount the worms or eggs. The preparation of the pad and the mounting of the worms or eggs is described in *Protocol 1*. If one-cell embryos are mounted carefully then all embryos should hatch. Worms should live several hours at least. In my hands, larvae are much more sensitive than embryos. For observation, larvae and adults can be anaesthetized using either 0.5–1% 1-phenoxy-2-propanol before mounting them on an agar pad containing 0.2% of the drug. Alternatively, the animals can be mounted on a pad containing 10 mM sodium azide (1).

Protocol 1. Mounting a specimen

Equipment and reagents

- A heater with a 7 × 9 cm aluminium block, which can be set at approximately 90°C. The block should have three holes, at one corner, to fit 10 ml test tubes. The remaining surface can be used to coat slides with polylysine for mounting eggs and worms for immunofluorescence.
- Erlenmeyer flask, 10 ml test tubes, Pasteur pipettes, small paintbrush, glass cutter, watchmaker glass, i.e. carved glass block, small alcohol or Bunsen burner, Parafilm, an eyelash glued to the tip of a toothpick, platinum spatula, scalpel

- Microwave oven
- Agar-agar (Difco)
- Vaseline
- Microscope slides, coverslips 18 × 18 mm or so
- Mouth pipette, 1,2 mm capillary glass (e.g. Science Products, cat. no. GB120–10) pulled out in a flame for collecting worms or eggs
- Stereo microscope
- Plate with worms

A. Preparing the agar pad

1. Boil 5 g of agar in 50 ml deionized water several times for a few seconds in the microwave oven. Distribute the agar in 15 test tubes. Heat the tubes again for a few seconds in the microwave oven. Seal the tubes with Parafilm. Store until use.

2. Briefly boil a tube of the agar over an alcohol or Bunsen burner. Put this tube, another one containing 5 ml of deionized water, and a third tube containing 5 ml of Vaseline into the heating block. Warm a Pasteur pipette by placing it into the tube containing the water. Hang the small paintbrush into the Vaseline. Stop it from touching the bottom of the tube by wrapping tape around the handle of the brush.

Protocol 1. *Continued*

3. Place three microscope slides side by side on the bench or on a piece of plastic. Cut a slide into two pieces. Keep one in hand. Use the warm Pasteur pipette to place a small drop of agar in the centre of the middle slide on the bench. Place a half slide on the drop and immediately press the agar into a very thin pad.

B. *Mounting the specimen*

1. Fill a watchmaker glass with a small amount of water. Collect your desired specimen from the plate of worms and place it into the watchmaker glass using a platinum spatula. Cut gravid hermaphrodites with the scalpel if you want to look at early eggs. Collect some specimens with the mouth pipette in as little water as possible.

2. Remove the slide from the top of the agar pad and place the pad under the stereo microscope. Use the mouth pipette to place the specimen in the centre of the pad. Withdraw most of the water with the mouth pipette. The amount of water left on the pad is critical as the specimen should not be covered with water. Avoid removing the specimen while removing the water by slowly sucking the water from the edge, change the position several times since the specimen moves slowly towards the capillary. Push the eggs or worms together into the centre of the slide using the eyelash.

3. Slowly place the coverslip on the pad. The droplet containing your specimen should touch the coverslip first. Slowly fill the space around the agar pad with water using the mouth pipette, but leave a small air bubble at one corner of the coverslip. Do not float the coverslip away from the pad.

4. Seal the coverslip with Vaseline using the small paintbrush—a thin, 3-mm wide strip will do. Do not put a huge amount of Vaseline on the coverslip or it will mess up your optics when pieces float around in the immersion oil.[a]

5. Mount the slide under the microscope.

[a] If there are only a few specimens on the slide it is helpful to mark the x and y position with a marker just outside the coverslip.

4.2 Setting up the microscope

The DIC, or Nomarski optics, is the most sensitive to set up, but it gives the most informative and beautiful images with the transparent *C. elegans* specimens. You should first check your microscope for the proper alignment of the optics (see *Protocol 2*) and then adapt the settings to your special preparation (see *Protocol 3*). These protocols are the summary of my personal experience. Normally the person delivering the microscope should do the alignment, but this person may not know what a good Nomarski image should look like. Once optimal resolution is achieved you can either take pictures directly with a classical photographic camera or use a video camera to capture a picture (see Section 7).

Protocol 2. Setting up a Nomarski microscope

Equipment
- A microscope equipped with Nomarski (DIC) optics
- Phase telescope

Method

1. Take off the light source and check the alignment of the filaments by projecting the light beam against the wall. Use the microscope manual for the specific procedure. Re-mount the light source.

2. Take out the polarizing filter and the condenser and turn the nosepiece for the objectives to an empty position. Replace one eyepiece with the phase telescope.[a] Make sure that all mirrors, lenses, and filters are absolutely clean to avoid scattering light in the microscope. Remove dust with an airbrush, and smears, stains, and old immersion oil with a cotton bud (Q-Tip) wetted with petrol ether.[b]

3. Close the aperture at the bottom of your microscope so that the remaining hole is right in the centre of your field of observation. Complain to your serviceman if this is not the case. Make sure that you can see the filaments of your lamp nicely centred in the aperture, open and close it to check this. Adjust the lamp again if this is not the case.

4. Re-mount the condenser and put it back up to the level of the stage. Set the condenser such that the Nomarski prism for the high-aperture lenses is in the light beam. Close the aperture of the condenser. Align the aperture of the light source and the condenser by moving the condenser with the appropriate screws. Make sure that the centres of the apertures are in the centre of your observation field; if not complain again.

5. Open the aperture of the condenser. Make sure that you see only one spot of the aperture of the light source shining through the top lens of the condenser (this is essential for good Nomarski). Complain again if you see several bright spots or reflections.[c]

6. Put the polarizer (first polarization filter) and heat filter back into the

---

Protocol 2. *Continued*

light beam. Position a slide, with the specimen mounted on an agar pad prepared as described in *Protocol 1*, on the stage and proceed as described in *Protocol 3*.

[a] The phase telescope permits focusing through your microscope. Normally it is used to align the rings of the phase-contrast optics, but it is also useful for spotting dirt in the light path of the microscope.

[b] Although it is nice to use expensive special lens paper, I find that normal tissue paper is also well suited to cleaning lenses.

[c] I have found that this adjustment is the most important step for generating a good Nomarski image. If you just see the original light beam centred in a slightly larger reflected light beam or no reflection at all the optics will be optimal. Nomarski optics requires that the polarized light which passes through a medium with the same refractive index is cancelled to darkness at the analyser (second polarization filter). If the light beam is reflected several times in the microscope this will be not the case, and the Nomarski image will be spoiled by the unwanted light contributions. This problem can sometimes be fixed by adjusting the position of the upper lens of the condenser. Ask the serviceman. I would not advise buying a microscope where this cannot be fixed.

Protocol 3. Setting up a specimen on the microscope

Equipment

- A microscope equipped with Nomarski (DIC) optics
- Mounted specimen on a microscope slide

Method

1. Mount the slide on the microscope stage. Focus, e.g. on the Vaseline sealant, using the 5 × lens of the microscope.[a] Close the aperture on the base of the microscope and focus the aperture by moving the condenser up or down. Search for your specimen by moving the stage. Focus and align the aperture to the centre of the observation field. Focus the aperture again with the condenser. Put a drop of immersion oil on the coverslip. Focus on your specimen.[b]

2. Change to the lens with high magnification. Move the stage slowly, usually upwards, until you see your specimen. Focus and align the aperture of the light source as described above.[c]

3. Turn the polarization filter (polarizer) to its darkest setting. Keep the light intensity rather low during this procedure. Turn the prism above the objective to its darkest position. Again turn the polarization filter to the darkest position and adjust the prism again. Make sure that the area around the aperture is very dark, while the specimen is very bright in the illuminated central part. Open the aperture of the light source to just out of the field of view. Turn the prism in either direction until you see a nice three-dimensional image of your worm or embryo.

(Examples of Nomarski pictures are shown in *Figure 1*.) The appearance may be better with the illumination from one side or the other. Check the general quality of the Nomarski as follows: move the slide to a position with no specimen, close the aperture of the light source, then go back to the darkest settings of the polarization filter and the Nomarski prism.[d]

[a] The aperture in the condenser must be completely open when using Nomarski optics. Sometimes I have observed sales representatives closing this aperture to increase the contrast of bad Nomarski optics. This is an inappropriate measure with which to improve the image, since resolution is lost by lowering the numerical aperture of the condenser.

[b] This is a very important step. Since the depth of focus is much higher with low magnification you may squash your specimen if you directly change to a lens (e.g. 100 x) with a high aperture. By focusing with the oil droplet you place the specimen safely far enough away for the higher magnification lens. Alternatively, move the stage a few millimetres down and then re-focus after changing the lens. However, I have squashed my specimens much more often with the latter method, by moving the stage up too hastily.

[c] Once this is set, I leave the condenser in position and, when I mount a new slide, place my specimen into the light spot seen with the 5 × lens. The optical axes of the low- and high-magnification lenses may vary significantly and a specimen centred with the low-magnification lens may be out of the field of view with the high-magnification lens.

[d] The area around the aperture should be almost black if you reduce the light intensity to a setting such that you can just see some light shining through the open centre of the aperture. If you do not see a very significant difference between the aperture and the area around the aperture you will not have a good Nomarski image. Ideally, the field of view should be very evenly illuminated. However, modern microscopes, with their very large field of view, can no longer be evenly illuminated. Therefore, it is sufficient if the centre, seen by the camera, shows a strong contrast between the illuminated and covered part of the field of view.

4.3 Analysing worms

You had heard that the worm is so simple, but now you have your favourite mutant under the microscope. What you see now is either a severely deformed larva or embryo (if you are working with a lethal mutation), or you cannot tell what is wrong at all, if your mutant is not lethal. In the first case a range of different stages of embryos should be mounted. Compare these embryos to normal embryos on a second slide to determine when development starts to deviate from normal. Embryogenesis is described in the seminal paper of Sulston and co-workers (2). A database containing a description of normal embryogenesis can be downloaded from a server (see Section 5.3.1). This database was created using a 4D-microscope (3). It links the lineage to a 4D-video and 3D-models of embryonic stages so that cell types and the position of cells can be inspected. When analysing mutants it is essential to follow early development from the one-cell stage up to about the 24-cell stage when gastrulation is initiated. It is my experience that even when the first cell divisions are aberrant, terminal stages may contain all the major tissues and look, apart from a generally disturbed morphology, quite healthy. A checklist may help if the first divisions are normal, where you score all the major structures like pharynx, hypodermis, intestine, body-wall muscle (indicated by a twitching of late embryonic stages), neuronal cells, and the germline pre-cursors. The general pattern of differentiation can be best analysed using late pre-morphogenetic and very early morphogenetic stages, i.e. late bean to 1.5-fold stages. Use the second slide showing various normal embryonic stages to flip back and forward between mutant and normal embryos. However, since most tissues are derived from several lineages you may find quantitative differences but not the complete lack of a tissue. Try to count the cells of the different tissues. At least pharyngeal, intestinal, and hypodermal cells can be estimated this way and strong alterations can be recognized. More exact numbers can be determined using immunochemical methods described in Chapter 9. *Protocols 4* and *5* describe two procedures for recognizing intest-inal and body-wall muscle in the light microscope. The correct number of body-wall muscles can only be determined using antibodies, since not all cells form the easily recognizable, spindle-like structures (4). In my experience, an embryonic phenotype can only reliably be further assessed by lineage analysis, preferably using a 4D-microscope as described in the next section.

The post-embryonic development of hermaphrodites and males was also described by Sulston and co-workers (5–9). This should be much easier to analyse than embryonic development, with the exception of the nervous system. Go step by step through the anatomy of normal animals and inspect the corresponding structures in the mutant animals. Behavioural studies may be very helpful for pinpointing a specific defect in the nervous system.

Protocol 4.   Scoring intestinal and body-wall muscle
differentiation with polarization microscopy

Equipment
• A microscope equipped with Nomarski (DIC) optics

Method

1. Mount your specimen under the microscope as described in *Protocols 1* and *3* for Nomarski microscopy.

2. Position the specimen properly and first inspect it using the DIC-optics to gain some orientation of the structures in the embryo or worm.

3. Change the setting of the filter wheel in the condenser to bright field. Remove the DIC prism above the objective. Put the polarization filter to its darkest position and reduce the light until the background is very dark.[a]

---

[a] Intestinal cells can now be recognized by their bright rhabditin granules which glitter like the stars at night. The integrity of body-wall muscle cells in larvae or adults is indicated by the presence of the regular array of muscle filaments (for further information see ref. 1).

Protocol 5. Scoring intestinal differentiation by gut-esterase staining

Equipment
- A microscope equipped with Nomarski (DIC) optics
- Heat block, Coplin jars, rubber policeman, 8-well Teflon-coated slides
- 0.1% poly-L-lysine solution
- 4% paraformaldehyde fixative
- 24 × 60 mm, coverslips
- A metal block that is sitting on dry ice

- 50 mM Hepes pH 7.0
- Staining solution: mix 250 μl of cold 4% pararosaniline (in 2.4 M HCl) and 250 μl of cold fresh $NaNO_2$. Leave on ice for 1 min. Add 10 ml 0.2 M Na_2HPO_4, 5 mM NaOH, and 2.5 mg α-naphthyl acetate dissolved in 250 μl dimethylformamide.
- 50% glycerol

Method

1. Prepare poly-L-lysine coated slides as follows. Heat an 8-well Teflon-coated slide to approx. 90°C on the heating block. Wet the rubber policeman with a drop of a freshly prepared 0.1% poly-L-lysine solution. Take the slide off the heating block and pull the rubber policeman across the slide three times—a thin film of the solution should evaporate immediately behind the rubber policeman. Heat the slide again for a few minutes. Allow the slide to cool down.

2. Place animals or embryos in distilled water into the wells. (Prepare a new slide using more poly-L-lysine than before if they do not immediately stick to the surface.) Remove the excess water so that only a thin film remains in the wells. Place a coverslip on the slide so that it hangs a few millimetres over the right side of the slide. Freeze the slide on a metal block that is sitting on dry ice. Flick the coverslip off and put the slide immediately into the fixative.[a]

3. Fix the specimen in ice-cold, fresh 4% paraformaldehyde for 10 min. Rinse the slide in 50 mM Hepes pH 7.0 for 10 min.

4. Immerse the slide in the staining solution.

5. Monitor the staining under a stereo microscope. Stop the staining reaction by washing in cold 50 mM Hepes pH 7.0.

6. Mount in 50% glycerol.

[a] Animals and embryos mounted like this can also be used for immunochemical and RNA *in-situ* analyses.

---

5. 4-dimensional microscopy

Embryos and worms have a considerable thickness. Therefore, development of the whole specimen cannot be followed using Nomarski optics with a lens of high-numerical aperture since the depth of focus is only a few microns. This prevents the documentation of all events by conventional videorecording. Hence, the seminal observation of the *C. elegans* lineage had to be carried out by directly observing and following the cell of interest by focusing through the embryo (2). As a consequence, labour-intensive lineage analysis could not really be used to directly analyse manipulated or mutant embryos. Thus, the precise mapping of all cell fates—which, in principle, should have allowed the alterations in aberrant embryos at a single-cell level to be determined—was not really possible in developmental studies.

These problems have now been solved with the invention of 4-dimensional microscopy. The time-lapse recording of multiple focal planes documenting events in three dimensions was pioneered by Minden and co-workers in 1989 (10) who recorded the behaviour of nuclei in *Drosophila* embryos. A high-resolution recording system based on an analog, laser video recorder was developed by White (11), which permitted the lineage analysis of *C. elegans* embryos (12). This set-up was further developed by my laboratory into a 4D-analysis system by improving the microscope itself and developing analysis software (3). The White laboratory has developed a digital recording system based on Macintosh computers (13).

5.1 What is 4D-microscopy useful for?

The function of a gene is defined by the phenotype caused by the lack of that gene's activity. Therefore the function of a gene can only be determined by a careful description of a given phenotype. There are different levels of phenotype description. The first level is the phenotype that can be described with no special effort by looking at worms with low magnification. Worms may be uncoordinated, dumpy, or show any other easily visible phenotype; a mutation may cause lethality or no apparent phenotype at all. These phenotypes are useful for an initial classification of the function of a gene; however, the next level is inevitably the description of the defect(s) at a cellular level. Essentially, a description at a cellular level can only be achieved by two methods: either by studying the expression of specific cellular markers using the various methods for studying gene expression, or by directly observing the structure and lineage of cells by light or electron microscopy. Electron microscopy can be very useful for describing the terminal phenotype. But only direct observation of cells during development permits the descent and the fate of cells to be connected. *C. elegans* offers, with the description of the complete lineage, a unique chance to map the pattern of developmental alterations caused by the lack of gene function. An example of how useful a thorough mapping of fate alterations may be is the analysis of the gene *lit-1* (14). Immunochemical analysis of the terminal phenotypes suggested that mutations in this gene essentially alter all lineages of the embryo, although a large variation in the specification of all major tissues did not suggest a specific function for the gene in cell-fate specification. Extensive lineage analyses, however, revealed that the gene functions in the binary diversification of anterior and posterior blastomere identities at least up to the 256-cell stage of the embryo, suggesting that 90% of all cells are specified by one general mechanism, which shares components with the Wnt pathway (14). The suitability of the worm for the determination of phenotypes to cellular and even subcellular resolution using light (4D) microscopy provides a great opportunity for the definition of gene function.

5.2 Technology

The principle of 4D-microscopy is to record, in a few seconds, enough focal planes so that all structures between the top and the bottom of a specimen can be seen. This procedure is repeated every 30 sec until the process of interest is finished. For *C. elegans* embryos, a recording of 25 focal planes, 1 µm apart, for approximately 7 h will document all stages of embryogenesis until the 1.5-fold embryo starts to move. After the recording the 20 000 or so images are ordered in 25 videos showing the different focal planes. These records can be replayed with the computer moving forward or backward through time and up or down through the focal planes, such that embryogenesis can be analysed by following the cells and the cell cleavages during development. If a cell moves out of a specific focal plane it can be taken up in the next focal plane. Terminal cell fates can be also determined. In mutant embryos in particular it is impossible to discriminate between the different kinds of neuronal or neuronal support cells. However, all major tissues, such as neurons, hypodermis, pharynx, and body-wall muscle, as well as programmed cell death, can be distinguished. A database called SIMI Biocell has been developed that permits the documentation of the analyses, and this will be described later.

When choosing a set-up for 4D-microscopy two features are crucial; the reliability with which the focal planes are recorded, i.e. the precision by which the position of the microscope stage can be controlled, and the optical resolution of the video record, as you can retrieve it. A significant problem in controlling the position of the stage is that some modern microscopes use a friction-based system for fine positioning. As a consequence, positional control is not absolute and will progressively deviate with the number of movements required. Therefore step motors in combination with such microscopes will give very unreliable stage control that requires constant manual refocusing during the recording. I solved this problem by developing a system that senses the absolute position of the stage during the recording (3). Very good resolution of the video record can be achieved using a Newvicon video camera in connection with a laser video recorder (LVR, Sony). Unfortunately this recorder is no longer produced by Sony, although used machines may be available on the market. Hopefully, in the near future digital video recording will be developed to a stage such that the large number of pictures required for 4D-microscopy can be stored and handled with the resolution required for this special purpose. The advice given above for choosing a normal micro-scope also applies in selecting a 4D-system. Do not buy an instrument with which you cannot yourself follow the lineage of an embryo even in the lowest focal planes. Do not trust theoretical calculations that the resolution cannot be better than that of the specific instrument you are offered. I have seen great differences in resolution between different systems (3, 11, 13). We have been able to lineage complete embryos even in the lowest focal planes (3), whereas others may claim that this is 'not possible' or difficult because of the mismatch between the aqueous index of refraction in the embryo and the refractive index of the lens/oil/coverslip.

Technical details of the different possible set-ups and possible advance-ments, for example in the technology of image storage, will not be discussed further here. Examples of a lineage analysis and recorded videos can be found on the Web at http://www.tu-bs.de/institute/genetik/schnabel/ce-home, and examples of digital recording at http://www.bocklabs.wisc.edu/imr/lm/lm.htm. These sites also offer links for advice.

5.3 Lineaging with a 4D-record

In principle, an embryo can be lineaged once a 4D-record has been acquired, with no special technology, by just following the cells and cell cleavages on the video screen and taking handwritten records. It helps to record the position of cells and cleavages together with the time and focal level on a transparency fixed to the screen. If you occasionally want to analyse some lineages this may be the method of choice. Systems offering digital picture storage are in-dependent of the actual 4D-microscope, and mean that embryos, which may be recorded in some other facility, can be analysed. Lineaging 'manually' is also something to consider when deciding whether a set-up is suited for your specific problem, with no financial obligation. Taking this simple approach initially is certainly better than not lineaging at all, and it may also help to overcome the initial concerns about using a new technology. At first, my laboratory lineaged using only handwritten records (12, 15–17). However, it is not really possible, or at the least very cumbersome, to extract all the information of a 4D-record this way.

5.3.1 A database for lineaging

To facilitate large-scale lineage analysis and for easier access to the wealth of data—such as cell cleavages, cell positions, and fates in the embryo—that are collected in a lineage analysis, I designed and commissioned new analysis software. The program is called Simi Biocell and is available from Simi Gmbh (D-85705 Unterschleißheim; http://www.simi.net). It is a Windows-based pro-gram that runs best on a computer equipped with a Pentium (or equivalent) processor. Images from a Sony LVD player, which offers a very high resolu-tion, are digitized with a frame grabber (Screen machine II from Fast Inc.)

and displayed on the computer screen. The program, however, is also equipped, or may be altered, to use digital records produced by any other recording system. Since digitizing the pictures lowers their quality significantly, at least in our hands, the pictures are also displayed in parallel on a high-resolution monitor when the LVD player is used. During lineage analysis, where the time and position in space of cells and cell cleavages are recorded, the software gradually builds up the lineage tree of the analysed embryo which is also displayed on the computer screen (see *Figure 2*). This tree can be

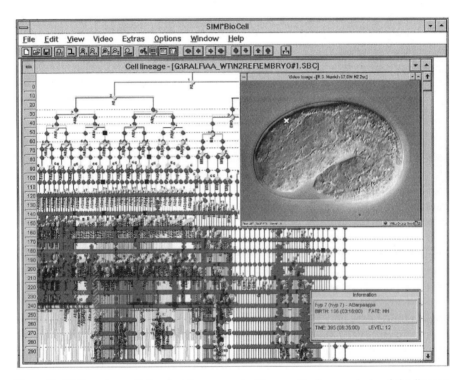

Figure 2. 4D-lineage analysis of embryogenesis. The result of an analysis of the lineage of a normal embryo using the SIMI Biocell database (3). The large window shows the complete lineage up to the pre-morphogenetic stage of the embryo (385 cells) and hypodermal, body-wall muscle, intestinal, and the germline lineages up to the 1.5-fold stage of embryogenesis (395 min). The dots in the lineage tree mark the points for which the positions of cells were stored. The reference lineage according to Sulston and co-workers (2) is visible in the background. The white cross in the video window shows the position of a hypodermal cell (hyp7 ABarpaappa) in focal level 12 at the end of the re-cording (395 min). These details are displayed in the 'information' window. The horizontal stippled lines force the observer to mark cells at specific time points. The positions of all cells can thus be displayed in three-dimensional models at these time points. This is a black and white representation of the windows that normally display the information in specific colour codes. As mentioned in the main text, a freeware version of the database can be downloaded from the World Wide Web.

compared to a reference tree (e. g. that of Sulston *et al.* (2)) shown in the background. While looking at the video, the position of a cell can be entered by just clicking with the mouse pointer on the cell nucleus in the window displaying the digitized image. This creates a mark at the position of the cell. In the lineage tree the time point of the mark is indicated by a dot on the appropriate branch of the tree. Any marked position or cell can be commented with information. By clicking on a dot in the lineage tree, the program shows the corresponding frame of the recording and marks the cell to identify it again; it also displays the name, fate, and any special information previously attached. A navigator can be used to guide lineaging towards a certain cell. A 'collision manager' checks the record for plausibility. A common error during lineaging is that cells in very close proximity are mixed up during movement between recorded focal planes. This inevitably leads to the assignment of the same position to cells from different origins. This apparent collision of two cells is detected by the program and a warning is given. In most cases the error can be easily corrected.

This system permits the comparison of different embryogenesis recordings by comparing the cleavage timing and the fate patterns in the embryos. The program allows 3-dimensional colour-coded representations of cells present at a given time. The 3D-model can be rotated in any direction to facilitate the comparison of embryos recorded in different orientations. It also permits comparison of the topological arrangements of embryos and investigation of the cell neighbourhood relationships. The movements of a cell or of groups of cells can be also visualized with the 3D-model. All information recorded in the database can be exported for further analysis. A freeware version of the software showing a complete lineage analysis of a normal embryo up to the pre-morphogenetic stage, including the corresponding 3D-video record, can be downloaded from a Web server (http://www.tu-bs.de/institute/genetik/schnabel/ce-home).

As already mentioned, this 4D-microscopy system was used to analyse the function of the gene *lit-1* (14) and for a new description of normal *C. elegans* embryogenesis (2, 3). It was also used for studies concerning the evolution of the buccal cavity of different nematodes (18). Priess' laboratory used the digital microscope developed by John White's laboratory for the lineage analysis of embryonic lethal mutants (19).

6. Confocal microscopy

You can use a confocal microscope if you do not have access to a sophisticated, normal fluorescence microscope. Confocal microscopes scan the viewing field with a laser microbeam and restrict the focal depth of the picture to reduce the unwanted contribution of scattered light. I used a confocal microscope to work on embryos for many years, but, in my opinion, normal microscopes, if used properly, can produce better pictures with the antibodies used for the analysis of worms. The worm or its embryos are thin and completely transparent structures so that the optical advantages of confocal microscopes are frequently not significant. Confocal microscopy may be advantageous in special situations, i.e. if resolution with a classical microscope is poor, because structures are brightly stained and very densely packed, for example in the nervous system, or because an antibody has a very strong background. When these instruments came on the market in the late 1980s they offered superior image analysis and storage software, which was very attractive and useful at that time. However, most of these features are now available with software packages that can be used with video cameras. If you are considering buying a confocal microscope, the advice given above still pertains: compare the performance of the instruments directly using the same specimen.

7. Documentation

As already mentioned, classical photography remains an alternative to general, digital-image documentation. In particular, the documentation of images requiring colour, like the β-galactosidase detection of gene expression, will often require normal photography since colour video cameras are still not generally used for microscopy. Cameras for fluorescence microscopy use grey-scale representation because of the higher sensitivity and resolution of black and white video cameras. Black and white pictures of coloured stains are not really satisfactory. Therefore, it may still be a good idea to use positive films for slides to document, for example *in-situ* hybridization, and then to generate prints from the slides. If equipped for high-quality printing of digital images, digitizing the slides and then processing the images for publication does allow perfect control over the colour balance. The colour of positives printed by photolaboratories may vary significantly from time to time, which can be a problem for the production of composite figures. Nevertheless, despite the wealth of methods for manipulating digital images, it is very important to produce a very good original, with the right brightness and contrast, irrespective of the method of image capture.

7.1 Photography
7.1.1 General considerations
Make sure that the specimen covers a large part of the picture, otherwise the resolution of the film is not properly used. However, compromise may be necessary here to achieve a sufficient focal depth to document all the structures you want to be represented in the pictures. It may be necessary to document several focal planes of your specimen, although one rarely sees a z-series through a specimen in publications. I have found that pictures taken with a lower magnification on a higher resolution film can give a much

better view of the structure in question than pictures taken at higher magnification.

As described below in more detail it is always advantageous to take a series of different exposures of a specimen. The measurement of the exposure time depends on the brightness of the actual structure you want to document, the background of your specimen, and the general background of your field of view. The relationship between these considerations may vary from specimen to specimen and from preparation to preparation. In principle, this problem can be minimized if spot measurement of light intensity, rather than measurement of the whole field, is used to determine the exposure time. However, I have found spot measurements much less reliable than measurements of the whole field. If possible, fluorescence pictures should be exposed so that the outline of the whole specimen can be seen on the picture.

7.1.2 Nomarski pictures

Taking Nomarski pictures using slide films with 100 ASA will give a very high resolution. The types of films available are changing all the time, but I use Ektachrome. The quality of the pictures may depend more on the laboratory developing your films than on the actual brand of film. A blue filter is recommended if daylight film is used. With regard to the temperature of the light source and the colour balance, it may be worth experimenting with different combinations of light intensities and exposure times. There are also the specific recommendations of microscope manufacturers to bear in mind. Under- and overexpose the pictures by a factor of two (correction –1 and +1, respectively) to start with in your first series. Depending on the result, optimize the corrections later. For black and white images I prefer the Kodak Technical Pan film. If this film is developed in HC110 developer diluted 1:10 for 6 min it will have a sensitivity of approximately 400 ASA and still have a very high resolution. When making prints it will be necessary to try different grades of paper to optimize the images. It may be worthwhile making the prints yourself, or to assist the technician in doing so, since only you know what the picture should really look like. Nomarski pictures show a strong shading from one side to the other. This can be removed by taking a longer exposure with shielding of the darker side of the picture, but only with considerable practice.

7.1.3 Fluorescence pictures

Since much less light is available in fluorescence microscopy I use 200 ASA films (Kodak Ektachrome) for slides and Technical Pan for black and white as described above. The quality of pictures, i.e. the resolution, improves significantly with an underexposure of a factor of two to four (correction –1, –2). In the case of a 200 ASA film this corresponds to exposures with 400 or 800 ASA, respectively. Try a more extensive series the first time, e.g. from 100 to 1600 ASA (correction +1 to –3). You can also produce double exposures, e.g.

using different fluorescence channels, by suppressing the winding of the camera after the first exposure. As a rule of thumb you should lower the exposure time of each image by a further factor of two. In double exposures with DAPI or Hoechst stains for DNA, the exposure time for a good representation of the nuclei should usually be corrected for by a factor of 8. It may be necessary to determine the correct exposure for these stains separately.

7.2 Collecting and processing digital pictures

Before considering the collection of digital pictures, a short comment on the manipulation of the captured images is warranted. Modern programs like Photoshop make it very easy to manipulate pictures to a degree inappropriate to scientific study. Make sure that you always save and keep an un-manipulated original of any picture you collect at your microscope. Any picture that is altered afterwards should be labelled as such in your records. Most manipulations, however, detract from the scientific quality of the image by removing original information, and therefore it is important for the original captured image to be of the best quality possible.

7.2.1 Nomarski pictures

I routinely use a Newvicon camera to collect Nomarski pictures. Pictures can be grabbed either directly from the microscope or from a Sony LVR using a high-quality frame grabber. I use a Pixel Pipeline in conjunction with a program called Biovision (Improvision, UK) running on a Macintosh computer. NIH Image also runs on a Macintosh, and there are now PC-based systems available. It is very important to set the intensity of the light source and sensitivity of the camera so that there are no completely black or white structures in your picture. The intensity of these structures, especially the completely white ones, cannot be manipulated later to be anything other than completely white. Collecting Nomarski pictures by averaging over several frames does not improve the picture. Using systems based on an analog camera gives a large picture, 27×18 cm, corresponding to 768×512 pixels with a resolution of 72 d.p.i. I process these pictures with Photoshop for printing. Usually these prints will be published in the size of a column in a journal, often 8 cm in width. Recalculating the resolution to 200–300 d.p.i. when reducing the picture by a factor of approximately three, maintains the resolution of the picture, at least when judged by eye. Higher resolutions make no sense, since printers do not usually print with a higher resolution. The same is true for the grey-scale representation: 8-bit corresponding to 256 grey values is enough. I am convinced that printers print less than that anyway, and I am told that my eye can only distinguish 16 grey values. When setting up a printer check the grey-value representation and adjust it if necessary. The printer should have a procedure to do so. Otherwise create a test file with a drawing program or Photoshop for the purpose. I am often very disappointed by the quality of digital printing that even high-end journals offer after I have spent many hours producing really top-quality images. My 'old' Sony UP-D7000E video printer with 172 d.p.i. usually prints pictures that can hardly be distinguished from those produced with the Technical Pan film recommended earlier. Therefore, again the advice before deciding to buy an extremely expensive high-resolution digital camera is to compare the final pictures on your desk with those acquired by the analog camera set-ups recommended by the NIH to run with NIH Image. When buying a printer, test your own pictures on different makes. Test pictures supplied by the manufacturers are specially selected to look good. When choosing a printer, the cost of each picture is also an important consideration.

7.2.2 Fluorescence pictures

Most considerations just mentioned for Nomarski images are also true for fluorescence images, except that either SIT (Silicon intensified target) or CCD cameras should be used. Because of the low levels of light the averaging of approximately 100 frames (either on the chip of the camera or with the frame grabber) greatly improves the pictures. Make sure you use the whole dynamic range of the camera, i.e. that the 256 grey values, or at least a large part of them, are really used. This could require dimming the mercury lamp or increasing the sensitivity of the camera. A setting that results in an occasional oversaturation of the brightest structure of your specimen will give a very good image after averaging. An inherent problem with using video cameras is that they have a small, linear dynamic range, whereas the human eye has a very large, non-linear dynamic range. As a consequence one can simultaneously resolve both very faint and very bright structures in a specimen, which video chips fail to do. The simplest solution is to collect several pictures with different intensities. Some modern high-resolution digital cameras also possess a very large dynamic range.

8. Electron microscopy

The EM was essential for analysing the anatomy of embryos and worms, and it was used to reconstruct the normal and mutant nervous systems (2, 20, 21). However, it appears that the electron microscope is not used much for the analysis of mutant phenotypes (22), although this is greatly recommended especially if cellular structures are affected. Transmission electron micrographs can be very useful for examining, for example, subcellular structures of membranes and the cuticle (22). Scanning electron microscopy was used to depict the morphogenesis of the embryo in very impressive pictures (23). An excellent description of electron microscopic techniques was written by Hall in 1995 (24). If you are not yourself an expert or you do not have an expert around, do not hesitate to contact D. Hall for advice. His laboratory supplies a service for the worm community.

References

1. Wood, W. B. (1988). *The nematode* Caenorhabditis elegans. Cold Spring Harbor Laboratory Press, NY.
2. Sulston, J. E., Schierenberg, E., White, J. G., and Thomson, J. N. (1983). *Dev. Biol.*, **100**, 64.
3. Schnabel, R., Hutter, H., Moerman, D. G., and Schnabel, H. (1997). *Dev. Biol.*, **184**, 234.
4. Schnabel, R. (1994). *Science*, **263**, 1449.
5. Sulston, J. E. (1976). *Philos. Trans. R. Soc. Lond. B. Bio. Sci.*, **275**, 287.
6. Sulston, J. E. and Horvitz, H. R. (1977). *Dev. Biol.*, **56**, 110.
7. Sulston, J. E., Albertson, D. G., and Thomson, J. N. (1980). *Dev. Biol.*, **78**, 542.
8. Sulston, J. E. and White, J. G. (1980). *Dev. Biol.*, **78**, 577.
9. Sulston, J. E. (1983). *Cold Spring Harbor Symposia on Quantitative Biology*, **48**, 443.
10. Minden, J. S., Agard, D. A., Sedat, J. W., and Alberts, B. M. (1989). *J. Cell Biol.*, **109**, 505.
11. Hird, S. and White, J. G. (1993). *J. Cell Biol.*, **121**, 1343.
12. Schnabel, R. (1991). *Curr. Opin. Gen. Dev.*, **1**, 179.
13. Thomas, C., DeVries, P., Hardin, J., and White, J. (1996). *Science*, **273**, 603.
14. Kaletta, T., Schnabel, H., and Schnabel, R. (1997). *Nature*, **390**, 294.
15. Hutter, H. and Schnabel, R. (1994). *Development* **120**, 2051.
16. Hutter, H. and Schnabel, R. (1995). *Development*, **121**, 1559.
17. Hutter, H. and Schnabel, R. (1995). *Development*, **121**, 3417.
18. Dolinski, C., Borgonie, G., Schnabel, R., and Baldwin, J. G. (1998). *Dev. Genes Evol.* 208, 495 (In press.)
19. Lin, R., Hill, R. J., and Priess, J. R. (1998). *Cell*, **92**, 229.
20. White, J. G., Southgate, E., Thomson, J. N., and Brenner, S. (1986). *Philos. Trans. R. Soc. Lond. B. Bio. Sci.*, **314**, 1.
21. White, J. G., Southgate, E., and Thomson, J. N. (1992). *Nature*, **355**, 838.
22. Bowerman, B., Eaton, B., and Priess, J. R. (1992). *Cell*, **68**, 1061.
23. Priess, J. R. and Hirsh, D. I. (1986). *Dev. Biol.*, **117**, 156.
24. Hall, D. H. (1995). In *Methods in cell biology* (ed. H. F. Epstein and D. C. Shakes), p. 395. Academic Press, San Diego, CA.

8

Neurobiology

JAMES H. THOMAS and SHAWN LOCKERY

1. Introduction

The nervous system of *C. elegans* is simple and uniquely well described. There are 302 neurons in the adult hermaphrodite that fall into 118 classes based on morphology and other characters (1). These neurons are interconnected by about 5000 chemical synapses and 600 gap junctions. The structure of each neuron is nearly invariant from animal to animal and the cell bodies are invariantly positioned so that each neuron can be identified *in vivo* (2). Despite its small size and simplicity, this nervous system regulates a wide variety of behaviours. These include several mechanosensory responses, many chemosensory responses, thermotaxis, complex responses to food, locomotion, feeding, egg-laying, defecation, and male mating. Neurons that control most of these behaviours have been identified, and in many cases they use identified transmitters and act in defined circuits.

In addition to its relative simplicity and the ease of genetic and molecular genetic analysis, *C. elegans* affords several other advantages for genetic analysis of behaviour. The nervous system of *C. elegans* is nearly dispensable for viability and hermaphrodite fertility. Only two neurons, CAN and M4 (J. Sulston, personal communication; (3)), are essential. Thus mutations with severe pleiotropic effects on the nervous system can be isolated as viable mutants (see, for example, ref. 4). Because of its optical clarity and invariant positioning of cell bodies, any neuron or group of neurons in live *C. elegans* can be killed with a laser microbeam, and the effects on behaviour studied (see, for example, ref. 3). Killing a neuron in the young larva results in its functional deletion from the nervous system, without measurable damage to the function of other neurons (see, for example, ref. 3, 5–7), thereby allowing assessment of the *in-vivo* behavioural function of any neuron or group of neurons. Finally, what has traditionally been the weakest feature of *C. elegans* neurobiology, the inability to record neuronal and muscle activity using electrophysiological methods, has now largely been solved.

This chapter has had to be selective. The amount of information available about the development and function of the *C. elegans* nervous system is so great as to preclude full coverage here. The reader is referred to several chapters in the recent book 'C. elegans *II*' (8) for comprehensive reviews of many aspects of *C. elegans* neurobiology.

2. Structure

2.1 Overview

The *C. elegans* nervous system consists largely of a nerve ring, a ventral nerve cord, a dorsal nerve cord, and a complex head sensory system (see *Figure 1*). The nerve ring is located in the head and contains almost all the interneurons together with axons from most sensory neurons; it is the brain of the worm. The ventral and dorsal nerve cords extend posteriorly from the nerve ring and run almost to the tail. They contain a variety of motor neurons, and the ventral cord contains processes from a number of sensory neurons and inter-neurons as well. Most nematode sensory neurons are in the head, presumably because this is the first part of the worm to contact a new environment during locomotion. In addition to these major regions of neuropil, there are many minor nerves that extend through the rest of the body.

The 118 classes of neurons in *C. elegans* have morphologies similar to those identified in other organisms, though generally simpler. A typical *C. elegans* neuron has a cell body with two long processes that are usually recognizable as a dendrite and an axon. The dendrite is usually unbranched and is either sensory or receives synaptic input from a small number of other neurons. There is little or no post-synaptic density visible by electron microscopy (EM) (1), a fact that complicates the ascertainment of synaptic connections, as discussed below. The axons are also usually unbranched and make synaptic output

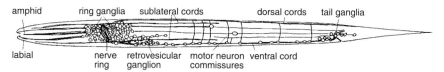

Figure 1. Overview of the nervous system. The cell bodies and major nerves of the young larva are shown in lateral view, head to the left. Labels indicate two of the major head sensory tracts, the major cell-body ganglia, and the major nerve cords that run the length of the body. All the nerves are situated near the outer surface of the animal, just beneath the hypodermis. The nerve ring runs all the way round the pharynx (not shown) and is by far the most complex neuropil in the animal. It contains the axons of most sensory neurons, axons and dendrites of nearly all interneurons, and the dendrites of the head motor neurons. These processes all run parallel to each other and course around the nerve ring to various extents. The structure of the adult hermaphrodite nervous system is very similar to the young larva, differing largely in the addition of ventral cord motor neurons and egg-laying motor neurons. The adult male has extensive additions of neurons in the tail region, which control mating. Ref. 1 and several chapters in ref. 8 provide much more extensive figures of the nervous system.

to a small number of other neurons or muscles. The pre-synaptic specialization is readily visible by EM and includes synaptic vesicles, darkening of the pre-synaptic plasma membrane, and usually a thickening of the axon that provides the necessary volume for vesicle storage. Nearly all synapses are *en passant*, meaning that the axon thickens and houses a pre-synaptic specialization, then narrows and extends further. Similarly, dendrites receive many synaptic inputs along their length without terminating or forming specialized branches or other structures. Based on structure and functional information, about half the classes of neurons are interneurons, about one-third are sensory neurons, and about one-quarter are motor neurons. (A few neurons appear to have mixed functions.)

2.2 Sensory anatomy

About 40 classes of *C. elegans* neurons appear to be sensory. Most of these are located in the head and project sensory dendrites toward the tip of the nose. Others are scattered in various locations throughout the body and in the tail. Based on their morphology, often corroborated by functional analysis, most sensory neurons are either chemosensory or mechanosensory. Most of the chemosensory neurons (all the ones with known functions) project sensory dendrites to the amphids (9), which are bilaterally symmetrical sensilla near the tip of the nose. Each contains the endings of 12 neurons. The sensory endings of mechanosensory neurons are structurally much more diverse and they are scattered throughout the body, though there is a notable enrichment in the nose (10). In addition to these sensory neurons, there is at least one thermosensory neuron (AFD, see ref. 11) and a few neurons with curious structures suggestive of special functions. For example, AQR and PQR have sensory cilia that project into the pseudocoelomic fluid (the nematode equivalent of blood) and both make synaptic output to interneurons. It seems likely that AQR and PQR report the composition of the pseudocoelomic fluid to the nervous system.

2.3 Motor anatomy

There are about 30 classes of motor neurons in *C. elegans*. These innervate body-wall, pharyngeal, egg-laying, and defecation muscles. The gross function of nearly all the major motor neurons is known. The structures of the neuromuscular junctions (NMJs) are diverse. The pharyngeal and egg-laying NMJs are at or near the ends of motor axons or branches that extend to the muscles. In contrast, most body-wall muscles and the defecation muscles send short neuron-like processes into nearby nerve cords where they form NMJs with their motor neurons. The innervation of the body-wall muscles of the head is even more unusual. These muscles send processes to a specialized region on the inner surface of the nerve ring where they receive complex motor input.

Innervation of body-wall muscles in *C. elegans* is more complex than for vertebrate skeletal muscle: most muscle cells receive synaptic input from three or more motor neurons, some of which are excitatory and others of which are inhibitory. This complexity may facilitate graded contraction of muscles using a small number of muscle cells, which results from a balance of graded excitation and inhibition by several classes of motor neurons.

2.4 Synapses

In a *tour de force* of electron microscopy, morphological criteria were used to tentatively identify all the synaptic connections among neurons and muscles in the *C. elegans* hermaphrodite (1). This effort resulted in what is often described as 'the complete wiring diagram' of *C. elegans*. While the morphological information is invaluable, there is good reason to think that this information is incomplete. One obvious difficulty is that most synapses are either excitatory or inhibitory, and electron microscopy gives no indication of which are which. Less obvious is the fact that most synapses release more than one transmitter, typically one small molecule transmitter and one or more neuroactive peptides. This combination is also common in *C. elegans* (1, 12), and different degrees and patterns of neuronal stimulation may result in the release of different mixtures of synaptic vesicle types. In short, chemical synapses are complex and not at all similar to 'wiring' (except rather complex circuits). Perhaps the worst difficulty is that the assignment of post-synaptic partners is problematic, because there is no discernible post-synaptic density and nearly all synapses are made *en passant* in regions of dense neuropil, without any glial cells surrounding the neurons. Because of these features, the site of vesicle release at most synapses is within 0.5 μM of several potential post-synaptic partners. White *et al.* (1) adopted the reasonable criterion that the real post-synaptic partner is the neuron (or two) that is directly apposed to the pre-synaptic darkening. However, the small extracellular volume and short distances to other neurons mean that several additional neurons might be strongly influenced by the same synapse. Conversely, if the neurons that are closest to the site of synaptic release do not express appropriate transmitter receptors then they will not be affected by the synapse, despite their proximity. Experiments in which neurons were killed suggest that many of the synaptic connections described by White *et al.* (1) are correct, but also that additional functional connections are likely to exist. Not until we possess a more complete description of the pre-synaptic transmitters and the distribution of post-synaptic receptors will we know just how accurate is the morphological assignment of synapses. A reasonable current rule of thumb is that a morphologically defined synaptic connection that is seen repeatedly is probably correctly assigned, but that occasional ones are suspect. Conversely, functional connections may exist that are not apparent morphologically because they occur between slightly separated processes.

2.5 Tools for analysing neuron structure

There are many methods for analysing both the gross and fine structure of neurons, such as sensory ending and synaptic structure. EM is the most difficult method, but it is still required for determining the finest structural detail, e.g. synaptic vesicle size and number. Whole-mount antibody staining is a simpler and effective method, provided, of course, that the appropriate antibodies are available (see Chapter 9). Considerable detail can be discerned with this method, including detailed process morphology and the positions of the varicosities associated with the pre-synaptic specialization. However, with the increasing availability of appropriate neuron-specific promoters, labelling with green-fluorescent protein (GFP) is quickly becoming the method of choice in most cases (13). This method is generally similar to antibody staining in resolution and specificity but is much easier technically and can be used with live animals, a critical advantage for many purposes such as screening for mutants with altered neuron structure. When expressed at sufficient levels without any localization determinants, GFP fills the entire structure of a neuron, facilitating cell identification and permitting detailed analysis of neuronal structure. An extraordinary recent example of this method is the visualization of live growth cones as they migrate (M. Bastiani and E. Jorgensen, personal communication). GFP is targeted appropriately when fused to protein determinants that confer specific subcellular localization, thereby permitting the analysis of specific aspects of neuronal structure such as the positions of synapses and sensory ending structure (see, for example, refs 14, 15). In addition to allowing neuron structure to be visualized, this method simplifies the determination of the subcellular localization of proteins of interest, although only in a transgenic context. Finally, double labelling is now possible with the advent of GFP variants having different emission spectra. This method has been used very little to date, but presumably it will be useful for determining the relative locations of different proteins or structures.

3. Genes

It has long been the hope that the *C. elegans* nervous system is a model for the function and development of nervous systems in general. The completion of the genome sequence has now made it clear that this hope is well founded. This section will briefly review the extraordinary conservation of molecular components of nervous systems of nematodes and vertebrates.

3.1 Neurotransmitters

Most classical neurotransmitter systems are present in *C. elegans*, including acetylcholine (ACh), glutamate, γ-aminobutyric acid (GABA), serotonin,

Table 1. Identified *C. elegans* neurotransmitters and probable neuronal distribution

Transmitter	Cells
Acetylcholine (ACh)	ALN, AS, DA, DB, HSN, IL2, M1, M2, M4, M5, MC, PLN, RIM, RMD, SAA, SAB, SDQ, SIA, SIB, SMB, SMD, URA, URB, VA, VB, VC, CA (male)
Dopamine (DA)	ADE, CEP, PDE, plus male R5A, R7A, R9A
GABA	AVL, DD, DVB, RIS, RME, VD
Glutamate (GLU)	ALM, ASH, M3, PLM, probably many others
Octopamine	Unknown
Serotonin (5-HT)	ADF, HSN, NSM, RIG, RIH, VC4, VC5, plus male CA, CP, R1A/B, R3A/B, and R9A/B
FMRFamide-related	AIA, AIM or AIY, ALA, AVA or AVE, AVK, DVB, HSN, IL1, I4, M1, OLL, PQR, PVT, RID, RIG, RMG, URB, VC, uv1 (non-neuronal)[a]

[a]All except IL1, OLL, URB, and PQR, which are based on a *flp-3::gfp* fusion (C. Li, pers. comm.), are based on antibody staining (12).
Adapted from the Lockery lab. Web page, URL: http://chinook.uoregon.edu/promotors.html

dopamine, and possibly glycine and noradrenaline (*Tables 1* and *2*). In addition, there are several demonstrated peptide transmitters and candidates for many others. In most cases, those transmitter receptor subtypes that have been identified in mammals are clearly recognizable in *C. elegans*. For example, mammalian glutamate receptors fall into three general families: AMPA/kainate, NMDA, and metabotropic. *C. elegans* has genes encoding at least seven AMPA/kainate subunits, two NMDA subunits, and three metabotropic subunits (our BLAST searches). Ionotropic ACh receptor subunits are even more diverse in *C. elegans*: there are over 35 genes encoding various subtypes of ionotropic ACh receptor subunits and there are putative metabotropic receptors as well (ref. 16, and my BLAST searches). In addition to the conservation of transmitters and their receptors, the mechanism of synaptic vesicle release is highly conserved (see, for example, ref. 17).

3.2 Neuropeptides

C. elegans appears to have a remarkable variety of neuropeptide transmitters, though little is yet known about their functions. There are more than 12 genes

Table 2. Neurons and muscles that probably express receptors for specific neurotransmitters

Transmitter	Cells	References
Acetylcholine (ACh)	All body-wall muscles, egg-laying muscle, pharyngeal muscle	24, 67, 89
GABA	All body-wall muscles, enteric muscles	28, 29
Glutamate (GLU)	AIB, AVA, AVB, AVD, AVE, AVG, PVC, RMD, SMD, URY	44, 95
Serotonin (5-HT)	Egg-laying muscles	23, 24

FLP-1

```
SAD PNFLRF
SQ  PNFLRF
ASGD PNFLRF
SD  PNFLRF
```

FLP-3

```
SPL GTMRF
TPL GTMRF
SAEPF GTMRF
NPL GTMRF
```

FLP-13

```
AADGA PLIRF
APEAS PLIRF
ASPSA PLIRF
SPSAV PLIRF
```

Figure 2. Neuropeptides encoded by three *flp* genes. Four representative peptide repeats are shown for each gene. The C-terminal segment is highly conserved within each gene (boxed). Each of these genes has several additional repeats with the same C-terminal identity (3 for *flp-1*, 5 for *flp-3*, and 4 for *flp-13*). A similar pattern holds for an additional 11 *flp* genes (18). The presence of diverse N-terminal segments and extensive third-base codon changes indicate that the peptide conservation within each gene is not an artefact of recent DNA duplications. It seems likely that each *flp* gene expresses its own distinctive peptide transmitter, presumably found in characteristic neurons and acting on distinct receptors.

that encode FMRFamide-related peptides, called the *flp* genes (*F*MRFamide-*l*ike *p*eptide; (18)). Each protein consists of a signal sequence followed by several repeats of a short peptide separated by two basic amino acids, where processing occurs to generate the mature peptide transmitter. The repeats in a given *flp* gene encode very similar peptides, but since the sequence of this peptide varies from gene to gene (see *Figure 2*) it seems likely that different *flp* peptides are expressed in different cells and that they are functionally distinct.

A second possible neuropeptide or neuroendocrine peptide family in *C. elegans* encodes a variety of insulin-related proteins (our unpublished analysis). Unlike classical neuropeptide transmitters, these proteins have a sequence typical of endocrine insulins: a signal sequence followed by B and A segments that are separated by an unconserved C segment. By analogy with other insulin relatives, this pro-protein is cleaved to liberate a B chain and an A chain that complex to form the mature ligand. There appear to be many such genes in *C. elegans*. Similar to the *flp* genes, these genes encode a diverse set of insulin-like proteins with the potential to act on many distinct receptors, suggesting an unprecedented diversity of insulin-related peptide signalling. One puzzle is that there appears to be only a single classical insulin receptor, DAF-2 (ref. 19 and our unpublished database searches). Either most of the insulin-like proteins are ligands for other unidentified receptors, or a wide range of insulin-like peptides all act on one receptor.

3.3 Voltage-gated ion channels

In addition to ligand-gated ion channels, voltage-gated ion channels and their relatives are also remarkably conserved in *C. elegans*. With the probable exception of the classical voltage-gated Na^+ channel, every class of these ion channels is found in *C. elegans*, often with several members. Even the lack of a voltage-gated Na^+ channel may not be particularly significant, since both in mammals and *C. elegans* the closely related voltage-gated Ca^{2+} channels can generate action potentials.

The most diverse family comprises the potassium channels, and we have analysed these in most detail. Most members of the potassium channel superfamily are voltage-gated, though a few are activated or inactivated, in part, by intracellular second messengers such as calcium. Almost every known potassium channel subfamily appears to have recognizable members in *C. elegans*. A dendrogram of some of these is shown in *Figure 3*. We have completed a similar analysis of inward rectifiers (*irk-1*, *irk-2*, and *irk-3*), the novel family of two-pore potassium channels (TWIKs, with about 50 members), BK potassium channels (n-SLO; A. Wei and L. Salkoff, personal communication), and SK potassium channels (C03F11.1, C53A5.5, and T02E1.8) with similar results. Much of this analysis has been independently corroborated (20). In most cases, the closest match to a particular *C. elegans* protein is a mammalian or insect protein rather than a related nematode protein, as can be seen in *Figure 3*. This fact indicates that the last common ancestor of mammals and nematodes already possessed this specific class of ion channel. The lower panel of *Figure 3* shows a detail of one part of this dendrogram: pairwise alignments between a short stretch of *C. elegans* EGL-2 and UNC-103 and each of these with the corresponding mammalian proteins. It is clear that, though these proteins are all closely related, the eag (EGL-2, d-eag, and r-eag) and the erg (UNC-103, seizure, and HERG) subfamilies had already diverged at the time of the first evolutionary split between mammals, insects, and nematodes.

3.4 Transmitter systems and ion channels as nuts and bolts

With few exceptions, it appears that the full complement of proteins that generate the complexity of nervous systems were already present in the common ancestor of nematodes, insects, and chordates (although mammals probably have more proteins of a given class—presumably as a result of the two rounds

(a)

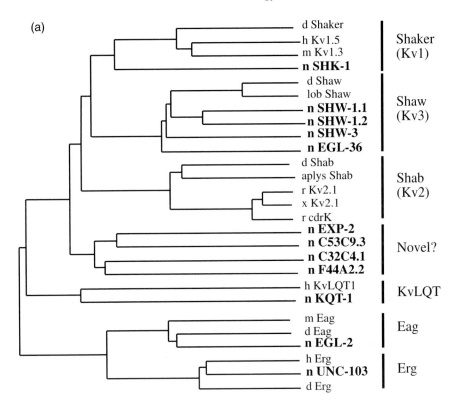

(b)

Figure 3. The K$^+$-channel family in *C. elegans*. The upper panel is a dendogram showing the sequence relatedness of a variety of *C. elegans* K$^+$ channels and their counterparts from insects and vertebrates. The lower panel shows pairwise alignments of a short region of the related K$^+$ channels EGL-2 and UNC-103 and mammalian homologues, with the number of identical amino acids at the right. EGL-2 and UNC-103 are each other's closest relatives in the nearly finished *C. elegans* genome. In the region shown and overall (see top panel), each is much more closely related to its mammalian counterpart, despite over a billion years of cumulative evolutionary divergence. EGL-2 and r-eag are members of the eag subfamily, and UNC-103 and h-erg are members of the erg subfamily. The last common ancestor of nematodes and mammals must have possessed two K$^+$ channels that would be readily recognizable as belonging to the eag and erg K$^+$-channel subfamilies.

of allodiploidization that occurred in the mammalian lineage). Despite this relatively close match in complexity at the protein level, insect and mammalian nervous systems have about 100 times and about 10^9 times as many neurons as *C. elegans*, respectively. This increase in neuron number is probably associated with a similar increase in processing complexity and memory capacity. We hypothesize that the dramatic discrepancy between organismal complexity and genetic complexity in the nervous system results largely from epigenetic phenomena. By analogy, a builder, supplied with an arbitrarily large number of nuts, bolts, and other building materials of (say) 500 types, can build a structure of arbitrary complexity. All 500 types of parts might be used to construct a few complex rooms and their interconnections, but the same 500 types of parts could build a million rooms and their interconnections merely by adding a few additional instructions about where to situate the various rooms and which to interconnect.

4. Behaviours

When a new mutant is identified, much can be initially inferred about it by testing for a variety of behavioural defects (*Table 3*), in a process not unlike a routine neurological examination in a human. In addition to this preliminary analysis, an assessment of the organism's *in-vivo* behavioural phenotype is a great strength that will continue to form a key part in the genetic analysis of the nervous system for the foreseeable future. A wide variety of *C. elegans* behaviours are under vigorous experimental attack in many laboratories. This section is a survey of the range of behaviours under study and specific assays for their analysis.

4.1 Locomotion

Locomotion in *C. elegans* is achieved by wriggling the body in sinusoidal waves. This motion is produced by 95 body-wall muscle cells with longitudinal fibres that are arranged in two dorsal and two ventral quadrants (2). The contraction of these muscles works on a hydrostatic 'skeleton', which is produced by maintaining positive pressure on the balloon-like flexible cuticle. Except in the nose, the body bends only dorsoventrally, as reflected in the equivalent motor innervation of the left and right members of the dorsal and ventral muscle quadrants (1). The broad brush-strokes of motor control of locomotion are established but many specifics are missing, including how bends propagate along the body, how the bends are kept in a smooth waveform, and how the frequency, amplitude, and rate of propagation of bends are controlled.

Body-wall muscles receive both excitatory and inhibitory motor input. Motor neurons that excite muscle on one side of the body also excite inhibitory motor neurons that act on the opposite side of the body (see *Figure 4a–d*),

Table 3. Summary of assays for major *C. elegans* behaviours

Behaviour	Assay	Reference
Locomotion	Qualitative inspection	65
	Radial dispersal rate	Thomas, unpublished
	Wave frequency	21
	Wave amplitude	21
Egg-laying	Egg stage	21
	Egg retention	22
	Egg-laying rate (liquid)	22
Defecation	Constipation	27
	Ethogram (direct inspection)	27
Chemoattraction	Point source gradient	7, 34
	Step gradient grid	Wicks and Plasterk, pers. comm.
Chemorepulsion	Ring crossing	40
	Gradient	Thomas, unpublished
	Step gradient grid	Wicks and Plasterk, pers. comm.
Thermotaxis	Radial gradient	43
	Linear gradient	43
Body-touch response	Plate tapping	96
	Eyelash	97
	Pick prod	98
Nose-touch response	Eyelash bumping	41
Pharyngeal pumping	Pump count	99
	Detailed inspection	100
	EPG	82
Male mating	Reproductive efficiency	63
	Direct inspection	62
Dauer formation	Uncrowded growth	49
	Crowding/starvation	49
	Pheromone response	50, 61
	Epistasis with *daf* genes	54

producing a bend in the body. There are several classes of cholinergic excitatory motor neurons (DA, DB, VA, VB, and probably AS) (*Table 4*), each of which is repeated several times along the length of the body. Similarly, there are several repeats of two classes of GABAergic inhibitory motor neurons (DD and VD). Based on their synaptic connections (1) and studies in which specific classes of neurons were killed (5), the DB and VB excitatory neurons generate forward movement (posteriorly propagating body waves) and the DA and VA excitatory neurons generate backward movement (anteriorly directed waves). The DD and VD inhibitory neurons are used as reciprocal inhibitors for both forward and backward movement.

The locomotory motor neurons are regulated by several pairs of cells called the command interneurons (5). The AVB and PVC command interneurons activate forward movement and the AVA, AVD, and probably AVE command interneurons activate backward movement (see *Figure 4a–d*). These interneurons receive extensive input from sensory circuits, both directly from sensory neurons (see *Figure 5*) and indirectly through amphid interneurons

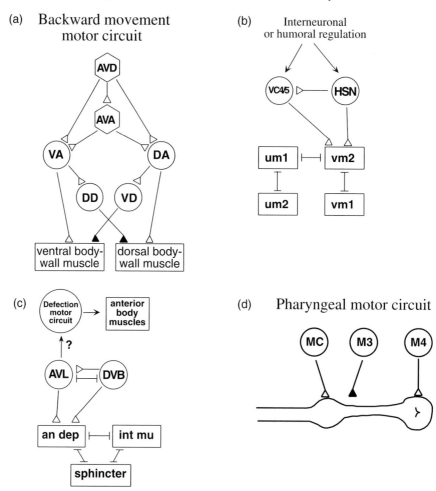

Figure 4. Motor circuits. Motor neurons are shown as circles and interneurons as hexagons. Putative excitatory synapses are shown as open triangles and inhibitory synapses are filled. Electrical junctions are shown by capped lines. Non-synaptic connections and putative synapses of unknown type are shown by arrows. (a) Backward movement circuit. The circuit controlling forward movement is very similar, except that the interneurons are AVB and PVC and the motor neurons VA and DA are replaced by VB, DB, and probably AS. The VD and DD reciprocal inhibitory motor neurons are shared by the two circuits, and their activity is controlled by the respective excitatory motor neurons. (b) Egg-laying circuit. (c) Defecation circuit. (d) Pharyngeal circuit.

and other nerve-ring interneurons. It is likely that the command interneurons serve to integrate complex sensory information and to direct appropriate locomotory responses. Each set of command interneurons makes synapses to the other set that are presumably inhibitory, ensuring that backward and forward movement are mutually exclusive states.

Table 4. Assigned functions of neurons, listed alphabetically

Neuron	Function	Reference
ADE, PDE	Regulate locomotion rate	Sawin and Horvitz, pers. comm.
ADF, ASI (ASG)	Repress dauer formation	6
ADL	Osmotic avoidance (tentative)	27
AIY, AIZ	Thermotaxis	11
ALM (AVM)	Light-touch response	5
AS	Locomotion motor neuron	5
ASE (ASK)	Soluble chemotaxis	34
ASH	Osmotic avoidance, nose touch	40, 41
ASJ	Activates dauer formation	51
AVA, AVD (AVE)	Backward command interneurons	5
AVB, PVC	Forward command interneurons	5
AVL, DVB	Enteric muscle motor neurons	28
AWA, AWC	Odorant chemotaxis	7
AWB (ADL)	Odorant avoidance	Bargmann, pers. comm.
CAN	Viability, possibly osmoregulatory	Sulston, pers. comm.
DA, VA	Backward locomotion motor neuron	5
DB, VB	Forward locomotion motor neuron	5
DD, VD	Reciprocal inhibitor motor neuron	5
FLP	Nose-touch	41
HSN	Egg-laying motor neuron	79
IL1V	Head withdrawal to touch	Kaplan, pers. comm.
PLM, PVM	Light-touch response	5
PVD	Harsh-touch response	98
RIA (RIB)	Movement control	11
RMD	Head withdrawal to touch	Kaplan and Horvitz, pers. comm.
M3	Pharyngeal motor neuron	82
M4	Pharyngeal motor neuron	3
MC	Pharyngeal motor neuron	89

Neurons listed in parentheses probably play minor roles in the same process, but evidence is less conclusive.

Assays for locomotion are diverse. A traditional qualitative method, sufficient for many purposes such as genetic mapping, is to observe the posture of an animal or to inspect briefly its spontaneous movement or movement in response to tapping the Petri plate. Because the range of movements in the wild-type nematode is limited and stereotyped, with practice even subtle defects in locomotion can be scored in this manner. For many purposes more specific and quantitative assays are necessary. Assays used include measuring the rate or amplitude of sinusoidal oscillation (see, for example, ref. 21) and measuring the rate of movement. The rate of movement can be approximated by several methods. These include (in order of increasing difficulty and resolution): picking several animals to the centre of a plate and measuring their relative dispersal over time (D. Reiner and J. Thomas, unpublished observations), picking a single animal to a plate and measuring its track length

Avoidance reflex circuits

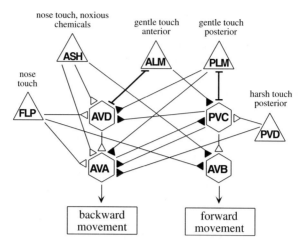

Figure 5. Avoidance reflex circuits. Sensory neurons are shown as triangles and their sensory modality is listed above them; see *Figure 4* legend for the remainder of the symbol key.

over time (J. Thomas, unpublished observations), or setting up a computer-controlled motorized stage and recording details of movement by tracking the image of a single worm (S. Lockery, unpublished observations).

4.2 Egg-laying

In the hermaphrodite, newly fertilized eggs are stored for some time in the uterus before being ejected into the environment. While in the uterus the eggs form a protective eggshell and begin the first several embryonic cell divisions. Egg-laying is driven by the simultaneous contraction of eight vulval muscles and eight uterine muscles that are interconnected by gap junctions. These muscles are innervated by two classes of motor neurons, the HSN and the VC neurons (see *Figure 4b*). When the HSN neurons are eliminated the animals become bloated with more and later stage eggs than normal (22, 23). Elimination of the VC neurons has subtler effects on egg-laying and enhances the egg-laying defect caused by loss of the HSN neurons (24). Quantitative analysis of egg-laying over time in individual animals revealed that most eggs are laid in brief bursts of about three or four eggs that are separated by an average of 20 min (24). The durations of the active state and the inactive state have a stochastic Poisson distribution. Mutant analysis suggests that serotonin from both the HSN and VC neurons induces the active state and that acetylcholine from both may actuate individual egg-laying muscle contractions during the active state (24).

Historically, egg-laying has been measured by picking animals from plates into microtitre wells filled with buffered salts, sometimes with drugs or neurotransmitters (see, for example, ref. 25), and counting eggs laid in the well over time. While this method remains useful for some purposes, interpretation of the results is complicated by various factors, including the non-physiological assay condition and mutant effects on the rate of egg production and the number of eggs stored in the uterus. Most recent studies have used assays of animals under normal plate-culture conditions. A good measure of the time from egg fertilization to egg-laying is the stage of embryogenesis of newly laid eggs (see, for example, ref. 26). By this assay, many mutants lay eggs much earlier or much later than the wild type. A similar assay would be to measure the stage of the oldest eggs in the uterus by mounting anaesthetized animals for Nomarski observation. An assay that measures a combination of the rate of egg production and egg-laying is to count the total number of eggs in the uterus. A measure of egg production that is relatively independent of the rate of egg-laying is to determine the rate of egg-laying over an extended time under normal growth conditions. In all these methods care should be taken in selecting the stage and physiological state of the animals to assay, as these affect egg-laying. Finally, an automated tracking and video recording method has been used to reveal details about the timing of egg-laying (24).

Protocol 1.   Egg-laying in microtitre wells

Equipment and reagents

- Well-cleaned, round-bottomed, plastic microtitre plate. (New plates may exude egg-laying inhibitors, so washing well and reusing plates is preferred.)
- M9 buffer (see Chapter 4, *Protocol 7*)

Method

1. Add 100 μl of M9 buffered salt solution to the desired number of wells. For drug assays dissolve the appropriate drug in M9 and use that solution.

2. Use a standard wormpick to put one animal into each well, use animals of appropriate stage and transfer as few bacteria as possible.[a]

3. After 60 or 90 min, count the number of eggs laid.[b]

[a] Staging the animals for assays is important! Typically you should pick animals with one row of eggs (about 15) in their gonad.
[b] The eggs tend to fall in toward the centre of a round-bottomed well and are easily seen. If you stagger picking the animals at the start of the assay you can stagger the counts correspondingly and assay a number of animals in one plate.

Protocol 2. Determining the developmental stage of newly laid
eggs

Equipment

• Dissecting microscope, and/or a high-   • Fresh seeded plate
  resolution dissecting microscope (e.g. Wild
  MZ12), or a compound microscope

Method

1. Pick between 10 and 20 adult animals of appropriate stage (see above)
   on to a fresh plate seeded with bacteria. Incubate for a short time
   (shorter times for more precise staging) and remove the parents.

2. For very approximate egg staging, simply observe the eggs on a
   regular dissecting microscope (good-quality optics are essential).

3. For more exact egg staging, view the plate on a high-resolution
   dissecting microscope such as the Wild MZ12 or transfer the eggs to
   slides and view on a compound microscope.

Note: A photo guide for later egg stages can be found in ref. 22.

4.3 Defecation

Defecation in the hermaphrodite occurs rhythmically about every 45 sec
under normal growth conditions. The muscle activity that mediates defecation
is surprisingly complex: first contraction and relaxation of the posterior body-
wall muscles in all four muscle quadrants (pBoc), then contraction of body-
wall muscles in the head in all four muscle quadrants (aBoc) co-ordinated
with the contraction of three types of specialized enteric muscle near the anus.
This stereotyped defecation motor programme (DMP) collects gut contents in
a pre-anal bolus and expels them. The specialized enteric muscles include two
intestinal muscles, one anal depressor muscle, and an anal sphincter. These
muscles are interconnected by gap junctions (1) and serve to squeeze the
posterior gut and open the anus (27); their coupled contractions are called the
expulsion muscle contraction (Exp or EMC). The enteric muscles are excited
by two redundant motor neurons, AVL and DVB (28), which use GABA as
their excitatory transmitter (29). It will be interesting to determine the
mechanism of this unusual excitatory role for GABA. AVL has a second role
in activating the anterior body-muscle contraction of defecation (28). This
role is unlikely to be directly as a motor neuron, since all parts of AVL are
distant from these muscles (1). In addition, GABA is not required for AVL to
activate anterior body muscles, suggesting that it uses a second transmitter in
this function.

Assaying defecation is readily accomplished by simply observing the muscle contractions (27). Since aBoc is the subtlest of the defecation steps and observing aBoc makes it harder to reliably score the EMC step, aBoc is often omitted from the observation. If determining the timing of defecation events is important, a simple data-collecting program is available that will record the exact time when computer keystrokes are entered (30) (with minor modifications, this program can be adapted for manually recording the timing of any behavioural feature). Most defecation mutations affect only some steps of the DMP, and the remaining steps can be used to indicate when a DMP has been activated. Thus the fraction of DMPs that execute each motor step can be determined with relative ease (27). Most mutants with defecation defects, especially those affecting the EMC step, are constipated (27). Their bloated gut lumen can be readily seen using a dissecting microscope, and it is particularly visible just posterior to the pharynx or just anterior of the anus. In addition, Con (*constipated*) mutants are starved (presumably due to reduced food throughput) and thus have the pale scrawny phenotype also seen in feeding mutants. Con mutants are easily identified and analysed genetically based on these phenotypes (27).

4.4 Pharyngeal pumping

Normal feeding consists of two pharyngeal motions, pumping and isthmus peristalsis (3, 31). A pump is a near-simultaneous contraction of the muscles of the corpus and terminal bulb (see *Figure 4d*), followed by a near-simultaneous relaxation. The contractile fibres of the pharyngeal muscles are radially oriented, so that contraction pulls the corpus lumen open and sucks liquid-borne bacteria in. During the subsequent relaxation, the liquid is expelled and most of the bacteria are trapped. About once every four pumps, isthmus peristalsis follows pump relaxation and transports trapped bacteria back to the terminal bulb. Contraction of the terminal bulb muscles breaks up bacteria by the movement of a specialized grinder structure and passes the debris back to the intestine (32).

Pharyngeal pumping is regulated by a distinct part of the nervous system that consists of 14 classes of neurons, directly interconnected with the rest of the nervous system through only one neuron class, RIP. When dissected from the rest of the worm, pharyngeal pumping is remarkably normal. Much of the complexity of the pharyngeal nervous system is not understood, but the major motor control is (see *Figure 4d*). Pump initiation is regulated by excitatory cholinergic input from the MC motor neurons, and the duration of each pump is regulated by inhibitory glutamatergic input from the M3 neurons. Isthmus peristalsis is controlled by excitatory input by the M4 motor neurons.

Mutants with reduced food ingestion due to pharyngeal defects are easily identified and analysed genetically on the basis of their starved appearance (33). Starved animals are smaller, thinner, and paler than well-fed animals and they mature more slowly. Direct observations of the pumping rate can be made using a dissecting microscope. Subtler defects in pumping motions can be analysed by mounting unanaesthetized animals with bacteria for observation through Nomarski optics. In a later section, the use of a simple electrophysiological method to discern electrical abnormalities in pharyngeal muscle will be described.

4.5 Chemotaxis

C. elegans moves up gradients of a variety of attractive stimuli and moves down gradients of other stimuli. The sensory neurons for many of these responses are identified. Most effort has gone into analysing their response to attractive odorants, which are sensed by the amphid neurons AWA and AWC. Together, these two classes of neurons respond to hundreds of distinct odorants (7). Based on cross-saturation studies, *C. elegans* can effectively discriminate between several classes of odorants, typified by benzaldehyde, butanone, and short-chain alcohols (sensed primarily by AWC), diacetyl and pyrazine (sensed primarily by AWA), and thiazoles (sensed by both). Though the molecular basis of this discrimination is not fully known, it is presumably based on batteries of distinct odorant receptors that are expressed in the appropriate neurons and separately linked to neuronal activity. Sensory neurons that mediate the attractive response to non-volatile chemicals (ASE and ASK (34)) and the avoidance of volatile repellents have also been identified (AWB, ADL, and ASH (35); C. Bargmann, personal communication). Whether a particular odorant will be attractive or repulsive appears to be determined by the sensory neurons where it is expressed. When the ODR-10 receptor for diacetyl, normally an attractive odorant, was misexpressed in the repellent-sensing AWB cell, the resulting animals acquired diacetyl repulsion (35).

Protocol 3. Assay for chemotaxis towards attractive odorants

Equipment and reagents

- Seeded NGM agar plates (see Chapter 4, *Protocol 2* and Section 3.3)
- S-basal buffer (see Chapter 4, *Protocol 3*)
- Chemotaxis agar plates: 10 cm Petri plates containing 10 ml of 1.6–2% agar, 5 mM KPi pH 6.0, 1 mM CaCl$_2$, 1 mM MgSO$_4$

- 1.5 ml microcentrifuge tubes
- Glass capillary or a Pipetman tip
- Drawn-out glass micropipette
- 1 M Na azide
- Freshly diluted odorant solution

Method

1. About 4 days before the assay, pick two to several adults (depending on the brood size of the parent) on to a seeded NGM growth plate to give enough progeny for about two assay plates.

2. Prepare the chemotaxis agar plates the day before the assay. If the air is humid, about an hour before the assay remove the lids to allow the surface of the agar to dry out a little.

---

3. With a felt-tipped pen, mark three spots on the back of each plate. Place two marks opposite each other and about 0.5 cm from the edge (odorant spots). Place the third mark on the midline between the first two and about 1 cm off-centre (worm spot). Use a marked plate lid as a guide to reproducibly place the marks.

4. About 15 min before the assay, pipette about 1.5 ml of S-basal buffer on to one or more of the growth plates. Gently swirl the buffer over the plate and use a Pasteur pipette to transfer it to a 1.5 ml micro-centrifuge tube. Allow the worms to settle on the bench for about 3 min, then aspirate most of the supernatant and similarly wash three times with S-basal buffer.[a] Use ddH$_2$O on the last wash to remove most of the salts.

5. During the washes, apply 1 μl of 1 M Na azide to each odorant spot on the assay plates.

6. Use a glass capillary or a Pipetman tip (with the tip clipped off to increase the bore) to transfer about 100 to 200 worms from the bottom of the last wash to the assay plate, using the smallest volume possible.

7. Leave the worms on the plate in the buffer droplet and apply 1 μl of the appropriate odorant solution or control to the marked spots on the assay plate.

8. Use a Kimwipe, or a drawn-out glass micropipette, to remove most of the liquid from the worm droplet to start the assay.

9. Put the lid quickly on to the plate and put the plate in a quiet spot (we put them under large inverted metal boxes to eliminate light and air currents). Ideally, use the same temperature for growth, washes, and assay (usually 20°C), but this is not essential.

10. Incubate for the desired time (usually 60 or 90 min) and then put each plate at 4°C in a single layer so that they chill rapidly, which stops the worms. At your leisure, remove the plates one at a time and count the worms at various positions on the plate, usually the two anaesthetic spots and the rest of the plate, leaving out worms that never moved away from the starting point (they are usually damaged or dead).

[a] Never let the worms sit in their pellet for long because it degrades their performance in the assay (they probably become anoxic).

Most assays for chemotaxis towards attractive cues have been made by setting up a relatively stable chemical gradient. For non-volatile attractants, a point source of attractant is allowed to diffuse in an agar-filled Petri plate prior to the assay. For volatile attractants, the gradient is formed in air just at the start of the assay by applying a point source of the attractant to the surface of the agar or to the plate lid. Most commonly in both cases, a non-volatile anaesthetic is applied to the attractant peak and to a diametrically opposed control spot (without attractant or with a different attractant) in order to trap animals that reach those points during the course of an assay. A count of those animals trapped at each anaesthetic spot, or neither, reveals information both about the efficiency of chemotaxis and the capacity to discriminate attractants.

Such assays have been used to identify and genetically characterize mutants that have specific chemotaxis defects (see, for example, ref. 36). Most such mutants are defective only in their response to certain compounds, indicating that they are competent in chemotaxis but defective in specific chemosensory transduction processes. The molecular identities and expression patterns of several genes identified in this manner support this view, and have contributed to an emerging picture of the mechanism of chemosensation, particularly of odorant response. Odorants are recognized by 7-pass receptors localized at the appropriate amphid sensory endings (15). Though not closely related in sequence, the existence of several large gene families of 7-pass receptors that are specifically expressed in chemosensory cells suggests that response to diverse odorants is mediated by a large number of receptors (35). The first receptor in any organism that was functionally assigned to a specific odorant was ODR-10, the receptor for diacetyl (15). Odorant receptors act through an intracellular pathway that includes trimeric G-proteins (37), cGMP (D. Birnby and J. Thomas, unpublished observations; C. Bargmann, personal communication), and cyclic nucleotide-gated cation channels (38, 39).

4.6 Chemical avoidance

Vigorous avoidance upon the abrupt exposure to noxious stimuli appears to be distinct from gradient chemotaxis. The chemical avoidance responses, together with noxious heat avoidance and harsh-touch avoidance, are arguably equivalent to nociception (pain sensation) in mammals. Noxious chemical stimuli that activate avoidance include high osmotic strength, heavy metal ions, detergents, and strong acid; all of these are toxic to *C. elegans*. When a moving animal contacts a region containing any one of these noxious stimuli, the animal abruptly reverses direction briefly, then makes a hairpin turn to move directly away from the noxious stimulus. As might be expected for noxious stimuli there is no evidence of adaptation to stimulus (J. Thomas, unpublished observations).

All analysed non-volatile, chemical avoidance responses are mediated by the amphids. Osmotic avoidance has been studied in most detail and is mediated predominantly by the ASH neurons (40). Interestingly, the ASH neurons also mediate part of the aversive response to nose touch (41) and to volatile repellents (42), indicating that they are polymodal nociceptors.

A variety of assays can be used to measure chemical avoidance responses.

One simple method is to print a ring of solution on to an agar surface free of food. Animals are then picked to the centre and the fraction of animals crossing the noxious ring is determined. With a carefully printed 2-cm ring containing about 10 μl of the applied solution, wild-type animals rarely leave the ring for the first 15 min of this assay, whereas strong amphid-defective mutants all cross within a few minutes. An alternative method, which may be preferable, is similar to the quadrant assay described for chemotaxis (adapted from Steve Wicks and Ron Plasterk, personal communication). This assay, though harder to set up, has several advantages, including the creation of a step gradient that is stable over the course of an assay.

Protocol 4.   Quadrant assay for attractants or repellents[a]

Equipment and reagents

- 2% agar solution containing the attractant (solution A) and an identical solution without attractant (solution B)
- 4-well, 9 cm Petri plates

Method

1. Melt and pour solution A into opposing quadrants of the 4-well plate and solution B into the other two quadrants. Slightly overfill the wells, but do not allow adjacent solutions to come into direct contact. Allow the agar to solidify and cool.

2. Immediately before the assay, apply a 1-mm layer of solution B across the entire surface of the plate and let it solidify briefly. (This layer bridges the gaps between wells.)

3. Prepare worms for assay (see *Protocol 3*). Place a 5–10 μl drop (containing 100–200 worms) in the centre of the assay plate. (The worms will begin to disperse after the drop is absorbed into the agar.)

4. At 10-min intervals, count the number of worms in quadrants containing solution A (N_A), and the number of worms in quadrants containing solution B (N_B). Compute the chemotaxis index (I), defined as: $I = (N_A - N_B)/(N_A + N_B)$.

[a] Thanks to Cori Bargmann, Steven Wicks, and Jon Pierce for contributing aspects of this protocol.

4.7 Thermotaxis

C. elegans has a finely tuned thermotaxis response (43). When placed on a surface with a thermal gradient, *C. elegans* displays a strong preference for a particular temperature. When the thermal gradient is radial, this preference often results in striking isothermal tracks on an assay plate. The preferred temperature matches the previous cultivation temperature over the range of 15°C to 25°C, and this preference changes gradually over a period of several hours when the cultivation temperature is shifted. There is also an element of plasticity in thermal preference, since animals that starve change their minds and avoid their cultivation temperature.

A neural circuit controlling thermotaxis has been described (11). A single amphid sensory neuron class, AFD, is required for isothermal tracking and normal temperature preference. The AFD neurons have a unique sensory ending that is composed of about 40 finger-like projections, inspiring their unofficial name 'finger cells'. It is likely that these fingers provide a large membrane surface for thermosensory transduction events. Steps downstream of the AFD sensory neurons are also partially understood. Thermotaxis seems to result from a balance between the activities of two amphid interneurons: AIY drives movement towards warmer temperatures and AIZ drives movement towards colder temperatures. AFD makes synaptic output to AIY, which in turn makes output to AIZ (see *Figure 6*). A simple model is that the synaptic activity of the finger cell is modulated by temperature, and that this activity regulates the relative contributions of AIY and AIZ to movement. AIY and AIZ may, in turn, function through a major integrating ring-interneuron class, RIA, though this part of the circuit is less well studied.

Thermotaxis assays have been performed in two ways. A simple, radial thermal gradient can be made by placing a vial of frozen glacial acetic acid in the centre of an agar plate. This assay is very easy, is suitable for assays of single animals, and reveals isothermal tracking nicely. A more easily quantitated assay that is usable for a large number of animals is based on a metal slab in which a stable linear thermal gradient is formed by immersing the two ends in water baths of different temperatures (43). A rectangular Petri plate

Thermotaxis circuit

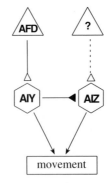

Figure 6. Thermotaxis circuit. In addition to identified parts of this circuit, Mori and Ohshima (11) inferred that there is at least one unidentified thermosensory neuron, here labelled '?'. See *Figures 4* and *5* for the symbol key.

with worms on the agar surface is placed on the slab for the assay. This assay is suitable for determining the average behaviour of many animals, but does not discern the ability of individuals to track isothermally.

Protocol 5. Radial gradient thermotaxis assay for single worms

Equipment and reagents
- Unseeded agar plates (see *Protocol 3*)
- Glacial acetic acid in scintillation vials

Method

1. A few hours before the assay put several scintillation vials filled with glacial acetic acid at 4 °C to freeze.

2. Several minutes before the assay, place a vial at the centre of an assay plate (on the plastic back) and place in a 25 °C incubator to set up the thermal gradient. Pipette a little water under the vial to establish good thermal contact with the plate back.

3. Pick one worm to assay on to a fresh unseeded plate and allow it to move away from the bacterial spot used in the transfer.

4. Pick up the worm in ddH$_2$O using a glass capillary, avoid transferring bacteria, and quickly put the worm about halfway towards the centre of the assay plate. Remove excess fluid, replace the vial, and put the plate back at 25 °C for up to an hour.[a]

5. Either trace the worm tracks with a pen, or place the plate on a piece of very high-contrast photographic film and expose it to a light flash to record the tracks. Use slightly drier assay plates if you have trouble seeing the tracks.

[a] You may have to replace the vial halfway through the assay if the glacial acetic acid melts too much.

4.8 Mechanosensory responses

C. elegans is endowed with a rich set of putative mechanosensory endings, but specific functions have been defined for only a few. When an animal runs into an obstacle with the tip of its nose it responds by briefly moving backwards (41). Two classes of sensory neurons, ASH and FLP, contribute to this response, and a third sensory neuron, OLQ, probably also contributes, albeit to a lesser extent. Both ASH and FLP make strong synaptic connections to AVA and AVD, the interneurons that activate backward locomotion. Thus, the nose-touch response is probably a simple reflex arc (44). There is an independent response to touch by an eyelash on the side of the body, called the Mec response. This process has been intensively studied and will be discussed below. In the absence of the Mec response, a response to a much harsher touch remains. This harsh-touch response is mediated by the PVD sensory neurons (45).

The Mec response to touch on the anterior part of the body is to move vigorously backwards, while the response to a posterior touch is to move forward, suggesting that these are also avoidance responses. Anterior touch is mediated by the ALM and AVM sensory neurons and posterior touch by the PLM and PVM sensory neurons. All these neurons have morphologically similar processes that run along the length of the body in their receptive fields, just under the hypodermis. A large set of Mec-defective mutants affect all these sensory neurons similarly, indicating that they share a genetic programme (45). The anterior touch cells activate backward movement through gap junctions to the backward command interneuron, AVD, and the posterior touch cells act similarly through the forward command interneuron, PVC. Thus, as with nose-touch, the primary Mec response is mediated by a simple reflex arc. Recent work has revealed plasticity and complexity in the touch circuit (46–48), indicating that even in this simple response there is considerable sophistication. In addition to their direct gap junctions to command interneurons, both the touch cells and associated neurons (such as BDU and LUA) make a variety of chemical synapses that are presumably important for these complexities.

4.9 Dauer formation

Dauer formation is included here as a behaviour because this developmental process is extensively regulated by the nervous system. The dauer is an alternative, third larval stage (49) that is induced by environmental stress, including crowding, temperature extremes, and reduced food availability (50). Response to these factors is mediated by several amphid sensory neurons, notably ASJ, ADF, and ASI. ADF and ASI act to repress dauer formation in the absence of dauer-inducing conditions (6), whereas ASJ activates dauer formation in response to dauer-inducing conditions (51). It is likely that additional sensory pathways mediate the response to the complex cues for dauer formation, including an unidentified thermosensory pathway (52). How information from these neurons is integrated and converted to a binary decision is not yet clear. Genes that regulate dauer formation have been ordered into a complex set of interacting pathways, largely by studying genetic interactions (52–55).

Relatively little is known about the mechanism by which ASJ controls dauer formation, except that it probably involves cGMP as a second messenger and cyclic nucleotide-gated cation channels, both of which also function in odorant response (56; D. Birnby and J. Thomas, unpublished observations). Repression of dauer formation by ADF and ASI is mediated by the secretion of a transforming growth factor-β (TGF-β) related peptide encoded by *daf-7* (51, 57). Response to the DAF-7 peptide is mediated by a heterodimeric serine–threonine kinase receptor and SMAD proteins, just as TGF-β and

BMP responses are in other organisms (58–60; T. Inoue and J. Thomas, unpublished observations). It is interesting to find a TGF-β, which are usually associated with developmental pattern formation, expressed in a neuron and regulated by environmental conditions (51, 57). A third pathway that regulates dauer formation involves an insulin receptor-like protein encoded by *daf-2* and other insulin-response genes (19), but how this pathway fits into nervous system function is currently unclear.

Dauer formation assays can be performed in a variety of ways. Two common assays are used simply to determine the frequency of dauer formation (vs. L3 formation) under non-dauer inducing conditions (i.e. normal uncrowded growth conditions; see ref. 49) and under strong dauer-inducing conditions produced by high temperature growth with limited food and exogenously added pheromone (50, 61). An advantage of these assays is that they are quantitative and they do not depend on normal locomotion or other behaviours.

4.10 Food responses

C. elegans responds behaviourally to the presence or absence of food in a plethora of ways. Most responses are easily rationalized as being adaptive for nutrient acquisition. For example, animals dramatically reduce their rate of pharyngeal pumping and defecation when food is removed. They also stop laying eggs, presumably preferring to provide their progeny with a food source, and they move more rapidly, presumably as part of active foraging for food. Surprisingly little progress has been made in understanding these responses, probably because they appear hard to perturb either by killing neurons or genetically. It is likely that the robustness of food responses results from the presence of multiple sensory pathways that act together to control this key aspect of worm behaviour.

4.11 Male mating

The most complex motor activity in *C. elegans* is performed by adult males in searching for and mating with hermaphrodites. Though hermaphrodites are surprisingly non-participatory, mating appears to be the primary goal of well-fed males. Recent work from the Sternberg laboratory has dissected several distinct motor activities of mating males, and has revealed that a variety of specific sensory cues direct these activities (62). The steps in mating include the response to contact, turning, vulva location, spicule insertion, and sperm transfer. The male's response to contacting a hermaphrodite is mediated by ray sensory neurons and involves clasping the hermaphrodite with his tail and running it along her body by moving backward. Turning is also mediated by ray sensory neurons and involves making a sharp ventral curl of his tail when the end of her body is reached, which permits him to run his tail back along her body, usually on her other side. Vulva location is mediated by sensilla just anterior and posterior to the cloaca (the hook and post-cloacal sensilla) and involves precise alignment of his cloaca with the vulval slit. Spicule insertion is mediated by sensory neurons in the spicule and involves the protraction of two spicules, which force the vulva open and anchor the male for sperm transfer. Soon thereafter, sperm transfer occurs, in part by hypercontraction of the anal sphincter, which opens the cloacal canal.

Mating efficiency can be easily measured by placing a fixed number of males on a mating plate with marked hermaphrodites and then counting cross-progeny production (see, for example, ref. 63). Using uncoordinated mutant hermaphrodites in such an assay allows measurement of the mating of those males that perform very poorly with non-uncoordinated hermaphrodites. More specific information can be obtained by directly observing attempted mating events and measuring the various visible steps in the process (62, 64).

4.12 Integrative functions

Relatively little is known about how interneurons integrate sensory information and modulate motor neurons and other effectors. The relative lack of information is probably because motor and sensory defects produce obvious behavioural defects, which are thus favoured in mutant screens. Nevertheless, some interneuronal functions are either known or inferred. Prime among these are the functions of the command interneurons for locomotion, which were discussed in the previous section. There is also some information about the roles of interneurons in directing thermotaxis (11) and chemotaxis (C. Bargmann, personal communication).

5. Pharmacology

The exposure of animals to exogenous neurotransmitters or other pharmacological agents has been of considerable value in the analysis of *C. elegans* behaviour (*Table 5*). Because these compounds often compromise locomotion, which is an essential part of many behavioural assays, they have been most useful in analysing motor output from the nervous system. We will briefly review the behavioural effects of such agents and their plausible interpretations. Then we will discuss the example of egg-laying in more detail, since this approach has been applied extensively in this case.

5.1 Overview of useful drugs

Most pharmacological agents that have been used in *C. elegans* affect various aspects of neurotransmission, often at the neuromuscular junction. As expected from the fact that ACh is an excitatory transmitter for body-wall muscles, the ionotropic ACh receptor-agonist levamisole causes severe body-wall muscle hypercontraction (65, 66). So far, three genes, identified through mutations conferring resistance to levamisole (66), encode subunits of an

Table 5. Neuroactive drug targets in *C. elegans*

Drug	Target in other organisms	Target in *C. elegans*	Genes
Aldicarb	Acetylcholine esterase	Acetylcholine esterase	*ace-1, ace-2, ace-3*
AMPA	Glutamate receptor	Glutamate receptor	*glr-1*
Fluoxetine	Serotonin reuptake	Serotonin reuptake	
Ivermectin	Glutamate-gated Cl⁻ channel	Glutamate-gated Cl⁻ channel	*avr-15*
Levamisole	Ionotropic ACh receptor	Ionotropic ACh receptor	*unc-29, unc-38, lev-1*
Muscimol	Ionotropic GABA receptor	Ionotropic GABA receptor	*unc-49*
Ryanodine	Ryanodine-gated Ca²⁺ channel	Ryanodine-gated Ca²⁺ channel	*unc-68*

ionotropic ACh receptor with properties similar to other ACh receptors (67). Similarly, the acetylcholine esterase inhibitor aldicarb causes body-wall muscle hypercontraction, though with a slower time course that probably reflects the time required for ACh to build up at the synapse (65). A large number of mutants have been isolated that are resistant to aldicarb but not to levamisole, suggesting that they affect the release or synthesis of ACh but not the post-synaptic response (68). Many of these genes are now known to affect general synaptic transmission (4, 17, 69–71). As expected from the fact that GABA is an inhibitory transmitter for body-wall muscles, the ionotropic GABA agonist muscimol causes flaccid paralysis of body-wall muscle (29). The *unc-49* mutants are resistant to muscimol, and it has recently been shown that *unc-49* encodes and ionotropic GABA-receptor subunits (B. Bamber and E. Jorgensen, personal communication). Dopamine also causes flaccid paralysis of body-wall muscles (72). Although it is not clear what the mechanism of this paralysis is, the result is consistent with a role for dopamine as a neuro-modulator. As expected from its agonist activity on muscle Ca^{2+} channels in mammals, ryanodine causes contraction of body-wall muscles in *C. elegans*. The gene *unc-68* encodes the nematode homologue of the ryanodine-activated Ca^{2+} channel and *unc-68* mutants are fully resistant to ryanodine (73). In contrast to vertebrate muscle, the *unc-68* channel is not essential for muscle contraction, suggesting that extracellular Ca^{2+} plays a greater role in muscle excitation. This may result simply from the much smaller size of *C. elegans* muscles, which increases their surface-to-volume ratio and places their muscle fibres close to the cell surface. Avermectins are agonists of glutamate-gated chloride channels, a receptor type that has, thus far, been described only in invertebrates (74, 75). Avermectins inhibit pharyngeal pumping at very low concentrations and cause flaccid paralysis at higher concentrations (76). Mutations in the avermectin-sensitive chloride channels can confer resistance to avermectins (77). AMPA, the agonist that defines the AMPA subtype of glutamate-gated cation channels, causes hyperactive foraging movements of the nose in *C. elegans* (44). Overexpression of one of the AMPA-type glutamate-receptor genes, *glr-1*, causes a similar phenotype (44), suggesting that AMPA has a similar target in nematodes and mammals. Finally, the invertebrate neurotransmitter octopamine blocks egg-laying in *C. elegans* (25), and there is at least one good match to an authentic insect octopamine receptor in the genome (C02D4.2; J. Thomas, unpublished observations).

5.2 Egg-laying pharmacology

Pharmacological approaches have been applied most extensively to egg-laying. This analysis has provided a key part of the evidence that the HSN neurons are excitatory serotonergic motor neurons for the egg-laying muscles. Exposure to serotonin induces egg-laying in animals suspended in liquid, a condition under which wild-type animals otherwise lay very few eggs. Similarly, serotonin-selective reuptake inhibitors (SSRIs), such as fluoxetine, also induce egg-laying (78); this result was first obtained with the less serotonin-selective reuptake inhibitors (22). Killing the HSN neurons abolishes egg-laying in response to SSRIs but not to serotonin. Finally, *cat-4* mutants, which lack detectable serotonin but possess HSN neurons that appear otherwise normal, lay eggs in response to serotonin but not fluoxetine (78). These data, together with the facts that the HSN neurons form neuromuscular junctions to egg-laying muscle (1) and contain serotonin (79), strongly support the model that the HSN neurons use serotonin directly to stimulate the egg-laying muscles to contract. There is recent evidence that the VC neurons may also contribute to this serotonin pathway (24).

5.3 Why are pharmacological targets so conserved?

Considering that current-day mammals and nematodes have diverged for an aggregate of over a billion years, the conservation of small molecular binding sites is perhaps surprising and begs an explanation. We speculate that pharmacologically active molecules have been designed (by nature or by man) to bind to key regulatory sites on their receptors. For example, muscimol is an ionotropic GABA-receptor agonist that binds at or near the GABA binding site. *Amanita muscaria* presumably evolved the capacity to synthesize muscimol to protect it from consumption by animals (perhaps insects). For such a strategy to be evolutionary stable, muscimol must bind to a site that is difficult to alter to non-binding, as this would confer drug resistance. One practical consequence of this conservation is that there is good reason to believe that generally useful answers will result from investigating the mechanism of action of neuroactive drugs in nematodes. The simplicity of *C. elegans* behaviour and nervous system combined with good genetics and a completed genome sequence should make such studies a new and powerful approach.

6. Electrophysiology

Until quite recently (80–82), the electrical properties of neurons and muscles in *C. elegans* could be inferred only from the genetic analysis or electro-physiology of other nematodes species (83, 84). This difficulty arose because the small size of *C. elegans* and the compactness of its nervous system, which make *C. elegans* so attractive for developmental and neuroanatomical studies, present a formidable challenge to electrophysiology. In addition, the body is protected by a tough, pressurized cuticle that explodes when dissected (85). These technical problems have been overcome, however, and we are now beginning to develop a picture of the electrophysiological principles of the *C. elegans* nervous system for the first time.

6.1 Electrophysiological methods

Electrophysiological recordings can be made from neurons and muscles in *C. elegans*. To record from a neuron, patch-clamp recordings are made from the cell body in a largely intact preparation (see *Figure 7A*). This preparation was inspired by the procedure used to make the first cell-attached, patch-clamp recordings from *C. elegans* neurons (D. Raizen, K. Breedlove, and L. Avery, personal communication). Animals (stages L1–adult) are placed on the surface of an agarose-coated coverslip that forms the floor of a glass recording chamber. Individual worms are immobilized with cyanoacrylate glue and immersed in physiological saline. Neurons are exposed for recording by nicking the cuticle of the head with a glass dissecting needle, forming a bouquet of 10–20 neuronal cell bodies. Using this preparation, it is possible to obtain most of the patch-clamp recording configurations, including cell-attached patches, excised patches, whole-cell recordings, and perforated patches (S. Lockery, unpublished observations; and see ref. 81). Neurons can also be exposed for recording from the midbody (S. Lockery, unpublished observations) and the tail (M. Goodman, personal communication). Thus, recordings can be made from almost any neuron in the animal. (See ref. 86 for a complete description of the method for patch-clamping *C. elegans* neurons).

Protocol 6. Preparing worms for electrophysiology[a]

Equipment and reagents[a]

- Agarose-coated coverslips
- Cyanoacrylate glue
- Physiological saline
- Patch pipette and recording equipment

Method

1. Glue 10–20 worms to a moist, agarose-coated coverslip using cyano-acrylate glue.

Protocol 6. *Continued*

2. Seal the coverslip to the recording chamber.

3. Immerse the worms in physiological saline.

4. Dissect a worm by nicking the cuticle near the target neuron. Several attempts on different worms may be required.

5. Seal the patch pipette to the neuron of choice.

6. Effect the required recording configuration and record.

[a] Details of this protocol are complex and evolving. A more up-to-date and detailed protocol can be found on the World Wide Web via the Lockery laboratory home page: `http://chinook.uoregon.edu/`

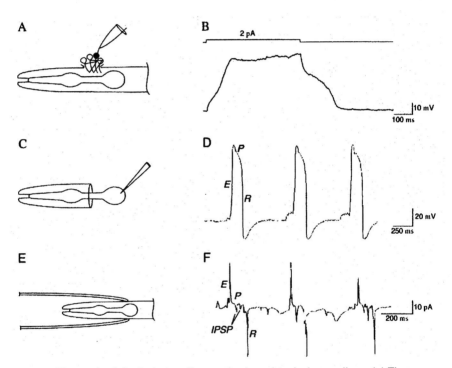

Figure 7. Electrophysiological recording methods and typical recordings. (a) The preparation for patch-clamp recording from neurons. Recordings are made from the cell body of neurons labelled with GFP (black cell). (b) The voltage response of the chemosensory neuron ASER to current injection (2 pA). (After Goodman *et al.* (81).) (c) The preparation for intracellular recording from muscles of the pharynx. (d) Pharyngeal-muscle action potentials. Each action potential has three phases (E, P, and R). (After Davis *et al.* (80).) (e) The preparation for recording the electropharyngeogram (EPG). (f) A typical EPG. Features corresponding to the E, P, and R phases of the action potential are indicated, as are two IPSPs. Note that the data in panels (d) and (f) are from different animals. (After Davis *et al.* (80).

After being exposed, neurons of particular classes can be identified by somatic expression of GFP. A permanent collection of strains in which a small number of neuron classes are labelled with GFP is rapidly accumulating, with more than 139 strains available already (`http://chinook.uoregon.edu/promotors.html`). In most strains, GFP expression is limited to about 12 neuron classes; and in a growing number of strains (currently 22), GFP expression is limited to a single neuron class. It is conceivable that in the future each neuron class will be identifiable by a specific GFP-labelled strain. In one unusual case, the chemosensory neuron class ASE, strains have been obtained in which either the left (ASEL) or right (ASER) member of the class expresses GFP (87). This surprising specificity allows one to return to precisely the same neuron in different animals. Consequently, much of what we currently know about neuronal physiology in *C. elegans* (see below) is based on recordings from the ASER.

Most of the work on *C. elegans* muscle electrophysiology has focused on muscle cells of the pharynx, since these are the biggest and most accessible muscle cells in *C. elegans*. Two different methods have been used. In the first method (80), intracellular recording, the pharynx is exposed by cutting the worm just posterior to the pharynx (see *Figure 7C*), and a conventional glass microelectrode is inserted into one of the muscle cells. In the second method (85), the electropharyngeogram (EPG), the head of the worm is drawn into a suction electrode and the currents that flow out of the worm's mouth during each pumping cycle are measured (see *Figure 7E*). The two methods give complementary results. In intracellular recordings, one observes 70–80-mV action potentials characterized by three phases (see *Figure 7D*): a rapid rising phase (E), a plateau phase (P) that typically lasts about 150 ms, and a brief repolarizing phase (R). Each phase of the action potential has a characteristic signal in the EPG recording, as indicated in *Figure 7F*. A preparation for making whole-cell, patch-clamp recordings from body-wall muscles has also been developed (J. Richmond, personal communication). Using this preparation it should be possible to study the genetics of synaptic transmission in *C. elegans*.

6.2 Neurons

The methods described above are being used to define the electrophysiological principles of the *C. elegans* nervous system (81), and how they are regulated by genes. Classical, brief action potentials cannot be elicited in ASER by current injection, nor are they observed to fire spontaneously (81). Rapid inward currents have not been detected in voltage-clamp recordings, consistent with the absence of genes for classical voltage-gated Na^+ channels in the *C. elegans* genomic sequence. However, an inward Ca^{2+} current, I_{Ca}, has been identified in ASER, raising the possibility of Ca^{2+}-dependent action potentials. I_{Ca} appears to underlie graded, regenerative responses detected in current-clamp recordings (see *Figure 7B*). The regenerative responses might play a role in signal amplification or transmission (see below). ASER also exhibits an inactivating, outward potassium current, I_K, activated by depolarization (81). I_K is opposed by I_{Ca} near the likely resting potential of the neuron, leading to a region of high input resistance, hence high sensitivity to sensory or synaptic inputs at the resting potential. Input resistance is reduced, however, at depolarized voltages where I_K becomes the dominant ionic current. It is thus likely that I_K contributes to both the maintenance and modulation of ASER's sensitivity. The ASE neurons are known to express four different potassium channel subunits (11, 39; J. Thomas, unpublished observations; L. Avery, personal communication). By recording from worms mutant for these subunits it should be possible to infer the role the channels play in sensory processing, integration, or other functions in ASER.

Recordings from other neurons suggest that many neurons in *C. elegans* may resemble ASER in the absence of Na^+-dependent action potentials, the presence of inactivating potassium currents, and the existence of a region of high input resistance near the resting potential. There is considerable variation, however, in the rate of potassium-current inactivation in different neurons. This is consistent with the large number of genes for potassium-channel subunits in the *C. elegans* genome (see *Figure 2*). Such differences could form part of the basis of functional diversity among the 118 neuron classes in *C. elegans*. Searching for functional differences between neurons should now be a simple matter of recording from other identified neurons, particularly those whose behavioural roles have been established by anatomical reconstruction and cell killing.

Signal transmission within neurons may involve passive and active mechanisms. ASE neurons are bipolar. Sensory transduction is thought to occur at the tip of one neurite, and synaptic outputs are located along the other neurite (1). In ASER, capacitance measurements and mathematical estimates of signal attenuation indicate that passive current spread is sufficient to send information from the transduction site to the synapses (81). Half of the neurons in *C. elegans* are similar to ASER in size and shape or have a single, short neurite (1). Thus, passive signal transmission is likely to be sufficient in many other *C. elegans* neurons.

Passive transmission is probably augmented by an active mechanism involving I_{Ca}. This follows from the fact that I_{Ca} is an inward current activated by depolarization, providing the possibility of regenerative feedback and, thereby, enhanced signal transmission. That positive feedback does indeed occur in ASER is suggested by the inflection point seen in the rising phase of the voltage response to a current injection (see *Figure 7B*). We do not yet know if other types of *C. elegans* neurons have regenerative inward currents sufficient for active signal propagation. However, genes for voltage-dependent calcium channels are known to be expressed in some other neurons (88).

6.3 Muscles

The excitation and contraction of pharyngeal muscle is triggered by a calcium-dependent action potential. The gene *egl-19* encodes the alpha-1 subunit of a homologue of vertebrate L-type voltage-activated Ca^{2+} channels (88). Gain-of-function mutations in *egl-19* cause long-lasting contractions, which are due to prolongation of the plateau phase of the action potential. Partial loss-of-function mutations in *egl-19* cause weak contractions, which are due to a reduction in the rate of rise of the muscle action potential. EPG recordings in the loss-of-function mutant are normal, emphasizing the greater resolving power of intracellular recordings. Nevertheless, EPG recordings are much easier and have the advantage that the animal can be saved for further analysis. The *egl-19* gene product is expressed in most muscle types in *C. elegans*, suggesting it may be the major Ca^{2+} channel in *C. elegans* muscles. This could be tested by recording from body-wall muscles.

6.4 Chemical synapses

The chemical synapse between the pharyngeal muscle motor neuron M3 and its target muscle is perhaps the most thoroughly characterized synapse in *C. elegans*, and several genes are known to regulate its electrophysiology (77, 82, 89). M3 is active during the plateau phase of the action potential, producing a burst of inhibitory post-synaptic potentials (IPSPs) that can be detected in EPG recordings from pharyngeal muscle (see *Figure 7F*). Variation in M3 IPSPs has been used to assess the effects on synaptic transmission of several genes, notably *avr-15*, which encodes a glutamate-gated chloride-channel subunit (77), and *snt-1*, which encodes a homologue of synaptotagmin (82), a protein important for Ca^{2+}-regulated, synaptic-vesicle release. M3 IPSPs are absent in *avr-15* null mutants, suggesting that glutamate is the neurotransmitter at this synapse. This view is supported by the fact that the plateau phase of the pharyngeal action potential can be shortened by a brief application of exogenous glutamate to the exposed pharynx (90). Null mutations of *snt-1* cause a behavioural phenotype consistent with a partial loss of Ca^{2+}-regulated synaptic function (4). M3 IPSPs in *snt-1* null mutants are smaller and more numerous than in the wild type, consistent with the behavioural phenotype of *snt-1* mutants and its homology to synaptotagmin.

6.5 Electrical synapses

Gap-junction proteins are necessary for the co-ordination of pharyngeal contractions (33). The *C. elegans* gene *eat-5* is believed to encode a structural component of gap junctions analogous to vertebrate connexins (91). Behavioural observations and EPG recordings indicate that muscle contractions in the anterior and posterior halves of the pharynx are synchronized. It is likely that synchrony is maintained by gap junctions between pharyngeal muscle groups, since they are dye-coupled and contraction synchrony continues after all pharyngeal neurons have been eliminated (3). In animals mutant for the *eat-5* gene, pharyngeal contractions exist but are unsynchronized and dye-coupling is lost, suggesting that *eat-5* is required for gap junction formation. The absence of electrical coupling could be tested directly by pairwise intracellular recordings of anterior and posterior muscle cells.

7. Future research

The description of the morphology and synaptic connectivity of all 302 neurons in *C. elegans* first raised the prospect of the comprehensive understanding of an entire nervous system. The addition of an arsenal of experimental techniques now places this goal within reach. The contributions of individual gene products can be assessed by mutational analysis, determination of complete expression patterns, and a host of other molecular genetic methods. The gross behavioural roles of individual neurons can be assessed by killing them in live animals. The largely conserved pharmacological targets permit the use of the extensive pharmacological tools developed for work in mammals and for the treatment of human disease. New electrophysiological techniques permit the assessment of detailed electrical aspects of neuron and muscle function. The small number of neurons increasingly permits analysis to consider the function of the nervous system as a whole, an advantage that is unlikely to be reproduced in any other nervous system in the foreseeable future. Finally, the complete genome sequence has clarified the astonishing degree of conservation of the molecular components throughout metazoans.

We predict that the most rapid progress in the near term will be made at the cellular level, where the electrophysiology of identifiable neurons can be combined with molecular genetics to exploit the large and growing collection of behavioural mutants. Information about the cellular properties of these neurons can then be integrated with electrophysiological recordings and mathematical modelling of the neural networks thought to be responsible for particular behaviours (92–94). The resulting synthesis should become the basis for the first comprehensive understanding of the behaviour of a metazoan.

References

1. White, J. G., *et al.* (1986). *Philos. Trans. R. Soc. Lond. B Biol. Sci.*, **314**, 1.
2. Sulston, J. E., *et al.* (1983). *Dev. Biol.*, **100**, 64.
3. Avery, L. and Horvitz, H. R. (1989). *Neuron*, **3**, 473.
4. Nonet, M. L., *et al.* (1993). *Cell*, **73**, 1291.
5. Chalfie, M., *et al.* (1985). *J. Neurosci.*, **5**, 956.
6. Bargmann, C. I. and Horvitz, H. R. (1991). *Science*, **251**, 1243.
7. Bargmann, C. I., *et al.* (1993). *Cell*, **74**, 515.

8. Riddle, D. L., *et al.*, (1997). C. elegans *II*. Cold Spring Harbor Laboratory Press, NY.
9. Ward, S., *et al.* (1975). *J. Comp. Neurol.*, **160**, 313.
10. Perkins, L. A., *et al.* (1986). *Dev. Biol.*, **117**, 456.
11. Mori, I. and Ohshima, Y. (1995). *Nature*, **376**, 344.
12. Schinkman, K. and Li, C. (1992). *J. Comp. Neurol.*, **316**, 251.
13. Chalfie, M., *et al.* (1994). *Science*, **263**, 802.
14. Jorgensen, E. M., *et al.* (1995). *Nature*, **378**, 196.
15. Sengupta, P., *et al.* (1996). *Cell*, **84**, 899.
16. Culotti, J. G. and Klein, W. L. (1983). *J. Neurosci.*, **3**, 359.
17. Nonet, M. L., *et al.* (1998). *J. Neurosci.*, **18**, 70.
18. Nelson, L. S., *et al.* (1998). *Mol. Brain Res.*, **58**, 103.
19. Kimura, K. D., *et al.* (1997). *Science*, **277**, 942.
20. Wei, A., *et al.* (1996). *Neuropharmacology*, **35**, 805.
21. Mendel, J. E., *et al.* (1995). *Science*, **267**, 1652.
22. Trent, C., *et al.* (1983). *Genetics*, **104**, 619.
23. Desai, C. and Horvitz, H. R. (1989). *Genetics*, **121**, 703.
24. Waggoner, L. E., *et al.* (1998). *Neuron*, **21**, 203.
25. Horvitz, H. R., *et al.* (1982). *Science*, **216**, 1012.
26. Johnstone, D. B., *et al.* (1997). *Neuron*, **19**, 151.
27. Thomas, J. H. (1990). *Genetics*, **124**, 855.
28. McIntire, S. L., *et al.* (1993). *Nature*, **364**, 334.
29. McIntire, S. L., *et al.* (1993). *Nature*, **364**, 337.
30. Liu, D. W. and Thomas, J. H. (1994). *J. Neurosci.*, **14**, 1953.
31. Albertson, D. G. and Thomson, J. N. (1976). *Philos. Trans. R. Soc. Lond. B Biol. Sci.*, **275**, 299.
32. Doncaster, C. C. (1962). *Nematologica*, **8**, 313.
33. Avery, L. (1993). *Genetics*, **133**, 897.
34. Bargmann, C. I. and Horvitz, H. R. (1991). *Neuron*, **7**, 729.
35. Troemel, E. R., *et al.* (1997). *Cell*, **91**, 161.
36. Culotti, J. G. and Russell, R. L. (1978). *Genetics*, **90**, 243.
37. Roayaie, K., *et al.* (1998). *Neuron*, **20**, 55.
38. Komatsu, H., *et al.* (1996). *Neuron*, **17**, 707.
39. Coburn, C. M. and Bargmann, C. I. (1996). *Neuron*, **17**, 695.
40. Bargmann, C. I., *et al.* (1990). *Cold Spring Harb. Symp. Quant. Biol.*, **55**, 529.
41. Kaplan, J. M. and Horvitz, H. R. (1993). *Proc. Natl Acad. Sci. USA*, **90**, 2227.
42. Troemel, E. R., *et al.* (1995). *Cell*, **83**, 207.
43. Hedgecock, E. M. and Russell, R. L. (1975). *Proc. Natl Acad. Sci. USA*, **72**, 4061.
44. Hart, A. C., *et al.* (1995). *Nature*, **378**, 82.
45. Chalfie, M. and Au, M. (1989). *Science*, **243**, 1027.
46. Wicks, S. R. and Rankin, C. H. (1995). *J. Neurosci.*, **15**, 2434.
47. Wicks, S. R. and Rankin, C. H. (1996). *Behav. Neurosci.* **110**, 840.
48. Wicks, S. R. and Rankin, C. H. (1996). *J Comp. Physiol A*, **179**, 675.
49. Cassada, R. C. and Russell, R. L. (1975). *Dev. Biol.*, **46**, 326.
50. Golden, J. W. and Riddle, D. L. (1984). *Dev. Biol.*, **102**, 368.
51. Schackwitz, W., *et al.* (1996). *Neuron*, **17**, 719.
52. Thomas, J. H., *et al.* (1993). *Genetics*, **134**, 1105.

53. Riddle, D. L., *et al.* (1981). *Nature*, **290**, 668.
54. Vowels, J. J. and Thomas, J. H. (1992). *Genetics*, **130**, 105.
55. Gottlieb, S. and Ruvkun, G. (1994). *Genetics*, **137**, 107.
56. Coburn, C. M., *et al.* (1998). *Development*, **125**, 249.
57. Ren, P., *et al.* (1996). *Science*, **274**, 1389.
58. Georgi, L. L., *et al.* (1990). *Cell*, **61**, 635.
59. Estevez, M. L., *et al.* (1993). *Nature*, **365**, 644.
60. Patterson, G. I., *et al.* (1997). *Genes Dev.*, **11**, 2679.
61. Golden, J. W. and Riddle, D. L. (1984). *Proc. Natl Acad. Sci. USA*, **81**, 819.
62. Liu, K. S. and Sternberg, P. W. (1995). *Neuron*, **14**, 79.
63. Hodgkin, J. (1983). *Genetics*, **103**, 43.
64. Loer, C. M. and Kenyon, C. J. (1993). *J. Neurosci.*, **13**, 5407.
65. Brenner, S. (1974). *Genetics*, **77**, 71.
66. Lewis, J. A., *et al.* (1980). *Genetics*, **95**, 905.
67. Fleming, J. T., *et al.* (1997). *J. Neurosci.*, **17**, 5843.
68. Nguyen, M., *et al.* (1995). *Genetics*, **140**, 527.
69. Miller, K. G., *et al.* (1996). *Proc. Natl Acad. Sci. USA*, **93**, 12593.
70. Iwasaki, K., *et al.* (1997). *Neuron*, **18**, 613.
71. Nonet, M. L., *et al.* (1997). *J. Neurosci.*, **17**, 8061.
72. Schafer, W. R. and Kenyon, C. J. (1995). *Nature*, **375**, 73.
73. Maryon, E. B., *et al.* (1996). *J Cell Biol.*, **134**, 885.
74. Cully, D. F., *et al.* (1994). *Nature*, **371**, 707.
75. Vassilatis, D. K., *et al.* (1997). *J. Mol. Evol.*, **44**, 501.
76. Avery, L. and Horvitz, H. R. (1990). *J. Exp. Zool.*, **253**, 263.
77. Dent, J. A., *et al.* (1997). *EMBO J.*, **16**, 5867.
78. Weinshenker, D., *et al.* (1995). *J. Neurosci.*, **15**, 6975.
79. Desai, C., *et al.* (1988). *Nature*, **336**, 638.
80. Davis, M. W., *et al.* (1995). *J. Neurosci.*, **15**, 8408.
81. Goodman, M. B., *et al.* (1998). *Neuron*, **20**, 763.
82. Raizen, D. M. and L. Avery. (1994). *Neuron*, **12**, 483.
83. Davis, R. E. and Stretton, A. O. (1989). *J. Neurosci.*, **9**, 415.
84. Jones, T. J., *et al.* (1991). *Rev. Nematol.*, **14**, 467.
85. Avery, L., *et al.* (1995). In C. elegans: *modern biological analysis of an organism*, (ed. H. F. Epstein and D. C. Shakes), p. 251. Academic Press, Orlando, FL.
86. Lockery, S. R. and Goodman, M. B. (1998). In *Methods in enzymology*, Vol. 293 (ed. P. M. Conn), p. 201. Academic Press, San Diego, CA.
87. Yu, S., *et al.* (1997). *Proc. Natl Acad. Sci. USA*, **94**, 3384.
88. Lee, R. V., *et al.* (1997). *EMBO J.* **16**, 6066.
89. Raizen, D. M., *et al.* (1995). *Genetics*, **141**, 1365.
90. Li, H., *et al.* (1997). *Proc. Natl Acad. Sci. USA*, **94**, 5912.
91. Starich, T. A., *et al.* (1996). *J. Cell Biol.*, **134**, 537.
92. Ferree, T. C. and Lockery, S. R. (1998). In *Computational neuroscience: trends in research* (ed. J. M. Bower), p. 373. Plenum, NY. (In press.)
93. Ferree, T. C. and Lockery, S. R. (1999). *J. Computational Neurosci.*, **6**, 263.
94. Ferree, T. C., *et al.* (1996). In *Advances in neural information processing systems 9* (ed. D. S. Touretzky, M. C. Mozer, and M. E. Hasselmo), p. 55. Morgan Kaufmann, San Mateo, CA.
95. Maricq, A. V., *et al.* (1995). *Nature*, **378**, 78.

96. Chiba, C. M. and Rankin, C. H. (1990). *J. Neurobiol.*, **21**, 543.
97. Chalfie, M. and Sulston, J. (1981). *Dev. Biol.*, **82**, 358.
98. Way, J. C. and Chalfie, M. (1989). *Genes Dev.*, **3**, 1823.
99. Avery, L. and Horvitz, H. R. (1987). *Cell*, **51**, 1071.
100. Avery, L. (1993). *J. Exp. Biol..*, **175**, 283.

9

Gene expression patterns

ANDREW MOUNSEY, LAURENT MOLIN, and IAN A. HOPE

1. Introduction

The first step after identifying a *C. elegans* gene of interest, either through a molecular approach or through sequence homology, is often to determine the gene's expression pattern. Expression pattern information is relatively easy to obtain and in combination with the sequence data can provide a valuable guide to subsequent strategies of investigation. For example, when and where a gene is expressed may direct the search for a mutant phenotype for that gene.

There are three commonly used approaches for studying gene expression in *C. elegans*: reporter-gene fusion techniques, mRNA *in-situ* hybridization, and immunofluorescent microscopy. Each has advantages and disadvantages. Reporter genes are simple to use, but leave concerns as to whether the expression pattern observed accurately reflects the expression of the native gene. A true reflection of a gene's expression can be obtained with mRNA *in-situ* hybridization, but for some weakly expressed genes the technique may be insufficiently sensitive. Immunofluorescence microscopy allows a gene's protein product to be located to subcellular structures, but this first requires the production of specific antibodies.

2. Reporter genes

Perhaps the most frequently used technique for the study of gene expression in *C. elegans* involves reporter genes. Plasmid expression vectors have been constructed containing reporter genes specifically designed for examining gene expression in *C. elegans*. A gene fusion is created between the *C. elegans* gene of interest and the reporter gene, by standard recombinant DNA techniques. The resulting recombinant plasmid is microinjected into *C. elegans* to generate stable transgenic lines (see Chapter 5). Expression of the reporter gene in the transformants is driven by regulatory elements of the *C. elegans* gene, and the distribution of the protein product of the reporter gene is discerned in the intact animal.

2.1 Selection of an expression vector

Andrew Fire and his laboratory have generated large sets of expression vectors (1). All these plasmids have a backbone (based on the pUC19 plasmid vector), containing the *E. coli amp*ʳ gene (encoding ampicillin resistance) and an origin of replication, that allows the facile cloning of the plasmid in bacteria. The hundreds of expression vectors available vary in the structure of the reporter gene that they contain. The reporter genes were assembled in a modular manner, however, and so selection of the appropriate expression vector is not as bewildering as it may at first appear.

Most of the reporter genes are derived from either the *lacZ* gene, encoding β-galactosidase, from *E. coli* (2) or the gene encoding green fluorescent protein (GFP) from the jellyfish *Aequorea victoria* (3). GFP is a fluorescent protein and direct observation of the distribution of this protein in live worms is facilitated by the transparency of the animal. Expression vectors are available encoding variant forms of GFP with different spectral properties (see Section 2.4). To examine the distribution of β-galactosidase, fixed animals are stained using a colourless soluble substrate that is cleaved to an insoluble coloured product at the sites of enzyme activity. The *E. coli* gene *uidA*, encoding the enzyme β-glucuronidase, has also been successfully used as a reporter gene in *C. elegans* (4).

All expression vectors have multiple cloning sites (MCSs), located both upstream and downstream of the reporter gene. The MCSs contain a number of unique restriction sites to facilitate the insertion of DNA fragments into the vectors. This means that not only can the upstream region of the *C. elegans* gene of interest be cloned in front of the reporter, but the downstream region of the gene, which may contain important regulatory elements (5), can also be inserted behind the reporter, if desired. Most of the expression vectors, however, have the 3′ region of the *C. elegans unc-54* gene in place downstream of the reporter, to provide the signals necessary for efficient processing of transcripts in *C. elegans*. This means that only the upstream region of the gene of interest need be cloned into the expression vector to drive expression of the reporter. The 5′ end of the reporter gene may either be fused to an exon of the *C. elegans* gene to generate a translational fusion or to the 5′ untranslated region of the *C. elegans* gene to generate a transcriptional fusion. The presence of MCSs flanking the reporter gene makes possible the alternative strategy of removal of the reporter gene from the expression vector for insertion into the gene of interest, in a plasmid or cosmid.

A nuclear localization signal (NLS) is encoded at the 5′ end of the reporter gene in some vectors. This signal can efficiently target β-galactosidase fusion proteins to the nucleus and may be helpful in identifying expressing cells. However, when the NLS resides at the junction of a fusion protein the degree of nuclear localization will depend on the particular context, and the *C. elegans* portion may take the fusion protein to other subcellular locations (6).

The nuclear localization of GFP fusion proteins is generally incomplete, with some cytoplasmic fluorescence usually being present.

The passing of β-galactosidase fusion proteins through a cell membrane, when they are secreted or membrane-bound, results in inactivation of the enzyme (2). To overcome this problem some vectors include a synthetic trans-membrane domain 5′ to the *lacZ* gene. The transmembrane domain contains a 'stop transfer' signal that should result in the β-galactosidase portion of the fusion protein remaining on the cytoplasmic side of the membrane and therefore retaining its activity.

Expression of transgenes is more efficient from spliced than unspliced transcripts (7, 8) and most reporter genes include a synthetic intron. The introduction of extra synthetic introns into reporter genes increased the level of reporter gene expression observed, and different expression vectors have different numbers of synthetic introns incorporated. However, the intron-rich *lacZ* vectors also give enhanced background staining, exacerbating one of the problems of using reporter genes.

Details of all the expression vectors referred to above and other more specialized vectors, together with further advice and procedures for acquiring the vectors are available on the Andrew Fire laboratory Website (`http://elegans.swmed.edu/Worm_labs/Fire/`).

2.2 Designing a reporter gene fusion

The rapid progress of the *C. elegans* genome sequencing project (see Chapter 2) means that the expression of virtually every gene in the genome is amenable to analysis by reporter genes. The almost complete cosmid coverage of the genome means that for any particular gene a cosmid containing that gene will be available. The sequence data in ACeDB (see Chapter 3) can be examined to identify a genomic DNA fragment suitable for fusion to a reporter gene. That gene fragment may be excised from the cosmid using restriction enzymes or amplified by PCR using the appropriate primers, either approach being based on the sequence data available. However, the following points ought to be taken into consideration when designing reporter gene fusions.

(a) *Size of genomic DNA fragment*: To obtain an accurate representation of the native gene's expression pattern all the gene's regulatory elements would need to be included. Therefore, as large a genomic DNA fragment as possible should be selected, preferably including the entire intergenic region between the gene of interest and the next upstream gene. Genomic DNA fragments of up to 10 kb can be inserted into *lacZ* expression vectors, although 5–7 kb fragments are easier to clone. Larger fragments may be cloned into GFP expression vectors because of the smaller size of the reporter gene.

(b) *Exon used in fusion*: If a gene's intron/exon structure is based purely on GENEFINDER predictions, and has not been confirmed by expressed sequence-tag (EST) data, cDNA analysis, or sequence homology, then it is important to be careful about which exon is used to create the fusion with the reporter gene. If the point of fusion is in an intron, no reporter gene expression will be obtained. Large exons with obvious open-reading frames should be preferred, with fusion in the middle of predicted exons rather than close to predicted intron/exon boundaries.

(c) *Inclusion of introns*: Many *C. elegans* introns contain elements regulating a gene's expression. In our experience fusion to the first predicted exon is more than twice as likely to yield no expression of the reporter compared to fusion to the second or a subsequent exon.

(d) *Polycistronic transcripts*: *C. elegans* is unusual amongst eukaryotes in generating polycistronic transcripts. Up to 25% of *C. elegans* genes are transcribed in this way. Unfortunately, there is no definitive way of recognizing possible polycistronic units from sequence data alone. However, if your gene of interest is less than 500 bp from a gene being transcribed from the same DNA strand, then it may be part of a polycistronic unit. In this case, regulatory elements for your gene of interest will be located upstream of the polycistronic unit.

2.3 Staining for *lacZ* expression

The *E. coli* gene *lacZ* encodes the enzyme β-galactosidase, which catalyses the cleavage of β-galactoside bonds. It was the first reporter to be used in *C.*

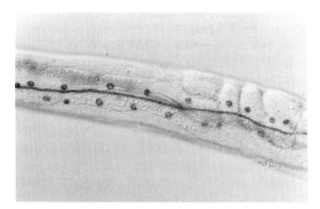

Figure 1. A mid-body view of a UL6 strain, adult hermaphrodite stained for β-galactosidase as described in *Protocol 1*. The UL6 strain was generated by transformation of the wild-type strain N2 with a *lacZ* reporter-gene fusion made using the promoter region of the *C. elegans* gene C17H12.4. The blue staining (dark in the figure) is observed in the excretory canals, which run laterally along the length of the worm, and in hypodermal nuclei either side of the lateral line. Colour images of this expression pattern can be seen on the World Wide Web at: http://www.personal.leeds.ac.uk/~acedb/Hope/html/c17h12.html.

elegans expression vectors and is still used extensively in gene expression studies in *C. elegans* and other organisms. The most sensitive assay for the detection of β-galactosidase *in situ* uses the chromogenic substrate X-Gal (5-bromo-4-chloro-indolyl-β-D-galactoside). Enzymatic cleavage of the colourless X-Gal results in the accumulation of a blue precipitate (see *Figure 1*). There are a number of methods for the histochemical detection of β-galactosidase in *C. elegans*. The following is a simple and robust method used routinely in our laboratory.

Protocol 1. Staining *C. elegans* for β-galactosidase activity

Equipment and reagents

- 5 cm seeded NGM plates containing worms (see Chapter 4, *Protocol 1* and Section 3.3 therein)
- M9 buffer: 3 g KH_2PO_4, 6 g Na_2HPO_4, 5 g NaCl, 1 ml 1 M $MgSO_4$, H_2O to 1 litre. Sterilize by autoclaving.
- 1.5 ml Eppendorf tubes
- 8-well, multiwell microscope slides (ICN Biomedicals, cat. no. 6040805) and 22 × 50 mm coverslips
- Methanol and acetone in separate Coplin jars at −20°C
- Redox buffer: 100 mM potassium ferricyanide, 100 mM potassium ferrocyanide. Keep stock at −20°C.

- Metal plate on dry ice
- Staining solution: For 1 ml of staining solution add the following in order: 620 μl ddH_2O, 250 μl 0.8 M sodium phosphate buffer pH 7.5, 1 μl 1 M $MgCl_2$, 4 μl 1% SDS, 100 μl 100 mM Redox buffer, 15 μl 5 mg/ml kanamycin (keep stock at −20°C), 2 μl 1 mg/ml DAPI (diamidinophenolindole) (keep stock wrapped in foil at 4°C), 8 μl 3% X-Gal in dimethyl formamide (keep at −20°C). Add the X-Gal last, and then mix quickly to avoid precipitation of the X-Gal. Staining solution can be kept for at least 2 days wrapped in foil at 4°C.
- Clear nail varnish

Method

1. Wash worms from 5 cm NGM plates, that have just cleared of bacteria, with 1 ml of dH_2O or M9 buffer into 1.5 ml Eppendorf tubes.

2. Allow the worms to settle under gravity for a few minutes or centrifuge briefly at 700 g (3000 r.p.m.) in a microcentrifuge.

3. Pipette 3 μl aliquots containing the worms from the base of the tube into the wells of 8-well microscope slides. Cover with a coverslip and place on a metal plate on dry ice for several minutes until the sample is frozen.

4. Flip off the coverslip with a razor blade and place the slide immediately in a Coplin jar filled with methanol, at −20°C, for 5 min. Avoid thawing of the sample.

5. Place the slide in acetone, at −20°C, for 5 min. Allow the excess methanol to drain off prior to placing it into acetone, but avoid thawing of the sample.

6. Drain off the excess acetone and allow the slides to air-dry dry at room temperature.

7. Add 25 μl of the staining solution to the slide. Carefully place a

Protocol 1. *Continued*

coverslip on to the slide so that the staining solution covers all 8 wells evenly, without trapping air bubbles.

8. Seal the edges of the coverslip with clear nail varnish.

9. Place the slides at 37 °C overnight or until staining is apparent.

Staining is observed with normal light microscopy using standard Nomarski optics (see Chapter 7). The inclusion of DAPI, which specifically stains DNA, allows the cell nuclei to be visualized and can aid in the determination of expression to specific cells. DAPI yields blue fluorescence when excited with UV light.

Where expression is weak, a 'super-sensitive' staining procedure (9) may be employed using a 30% stock solution of X-Gal in dimethylformamide (DMF). This solution is very unstable, oxidizing within hours at room temperature, although it may be kept for up to a week at –70 °C. The staining procedure is essentially as in *Protocol 1*, with the following modifications. The staining mix is heated to 65 °C before adding 10 μl of the 30% X-Gal solution with rapid mixing. The staining mix is then cooled to 45 °C before applying to slides pre-warmed to 42 °C to prevent precipitation of the X-Gal. Slides are incubated at 42 °C overnight or until staining is apparent.

Embryonic expression is easier to observe in embryos that have been removed from adults. There are two methods that can be used to achieve this, but both require the use of polylysine-coated slides as embryos stick poorly to uncoated slides. Polylysine-coated slides can be prepared using *Protocol 2*. After worms have been placed on the slide (see *Protocol 1*, step 3), gentle pressure is applied above each well which should release embryos from most adults. Alternatively, embryos can be prepared using the hypochlorite method outlined in Chapter 4, *Protocol 5* with thorough washing of the eggs in M9 buffer before placing 3 μl aliquots into the wells of coated slides.

Protocol 2. Preparation of polylysine-coated slides

Equipment and reagents

- 8-well, multiwell microscope slides (ICN Biomedicals, cat. no. 6040805)
- Poly-L-lysine. Store dessicated at –20 °C
- Heating block
- Rubber policeman

Method

1. Prepare a 1 mg/ml solution of polylysine in water. Dissolve by vortex-ing. Prepare on the day of use.

2. Thoroughly clean the slides with laboratory detergent to remove any traces of grease.

3. Heat the slides to 80 °C on a heating block.
4. Dip a rubber policeman into a drop of polylysine solution and evenly spread the liquid across the hot slide. Repeat three times.
5. Heat the slides for another minute at 80 °C.
6. Cool the slide to room temperature before use.

2.4 Examination of GFP expression

Green-fluorescent protein (GFP) is a protein of 238 amino acids that was isolated from the jellyfish *Aequorea victoria*. The wild-type protein is fluorescent, absorbing blue light at 395 nm (with a minor peak at 470 nm) and emitting green light at 509 nm (with a shoulder at 540 nm) (10). Used as a reporter of gene expression, GFP has the distinct advantage that it does not require the application of exogenous substrates. Thus it is possible to monitor gene expression in living worms and to directly follow the spatiotemporal variation in gene expression. Under the control of an appropriate promoter, GFP can be used as a vital marker to follow the outgrowth and migration of cells *in situ*. Furthermore, screens for mutations that alter specific gene-expression patterns are considerably simplified by the use of GFP as a reporter. As gene expression is examined in viable worms, those displaying altered gene-expression patterns can be recovered and their progeny isolated for further analysis.

Although GFP has many advantages, there are some disadvantages. GFP lacks the amplification intrinsic to enzymatic reporter systems. Each GFP molecule represents a single fluorophore, and relatively high levels of GFP expression (perhaps 10^6 molecules per cell) may be necessary for bright signals. For some *C. elegans* genes, expression of GFP fusions is difficult to detect compared to *lacZ* equivalents. There can be a considerable time lag between the expression of the GFP fusion and the onset of fluorescence. The formation of the GFP fluorophore is a two-step process involving cyclization of Ser65, Tyr66, and Gly67 followed by oxidation of the tyrosine (for a review see ref. 11). Measured time constants for the final oxidation step have been 2–4 h for wild-type GFP. This delay in the acquisition of fluorescence can be a particular problem for the examination of gene expression in early or mid-stage embryos, when embryonic development is rapid. However, there are GFP variants with improved properties. In particular, GFP variants S65T and S65C (change of serine-65 to threonine and cysteine, respectively) give a more intense fluorescent signal and a fourfold faster oxidation step in fluorophore formation (12, 13). These mutations also alter the spectral properties of GFP, shifting the excitation and emission maxima (to 479 nm and 507 nm, respectively, for S65C and to 488 nm and 511 nm, respectively, for S65T). Vectors containing the mutant forms of GFP are available in the Andrew Fire vector kit.

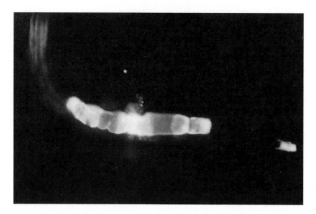

Figure 2. A mid-body view of a UL594 adult hermaphrodite examined as described in *Protocol 3*. The UL594 strain was generated by transformation of the wild-type strain N2 with a GFP reporter-gene fusion made using the *C. elegans* gene F18G5.2. The green fluorescence (white in the figure) is particularly obvious in the wall of the uterus (centre) and in the rectal region (right). Images of a strain transformed with a F18G5.2::*lacZ* fusion can be seen on the World Wide Web at `http://www.personal.leeds.ac.uk/~acedb/ Hope/html/f18g5.html`

   Live, actively swimming worms can be viewed directly for GFP fluorescence in a drop of water or M9 buffer. However, for detailed analysis and recording of expression patterns the worms should be anaesthetized and mounted using the method described in *Protocol 3*. Sodium azide is the preferred anaesthetic as phenoxypropanol (the other commonly used worm anaesthetic) quenches GFP fluorescence. GFP fluorescence can be viewed using the standard FITC filter set (see *Figure 2*). It should be noted that the gut granules (see Chapter 7, *Protocol 4*), in the intestinal cells, autofluoresce, but appear yellow in comparison to standard GFP and are therefore distinguishable.

Protocol 3.   Viewing worms for GFP fluorescence

Equipment and reagents

- Plain glass slides and coverslips
- Agar
- Platinum wire wormpick

- 10 mM sodium azide in M9 buffer (see *Protocol 1*)

Method

1. Place a drop of melted 5% agar in M9 buffer (containing 10 mM sodium azide) on to a plain glass slide.

2. Place a second glass slide over the agar drop so that the agar is spread into a thin even layer about the thickness of adhesive tape. The second

---

slide can be supported on 'spacer' slides raised by one layer of adhesive tape to ensure the correct thickness of agar.

3. Separate the slides to leave the agar pad adhered to one of the slides.
4. Place 1–2 μl of M9 buffer on the agar pad.
5. With a platinum wire wormpick, pick 5–15 worms into the drop of M9 buffer.
6. Immediately place a coverslip on to the agar pad. The worms should stop moving within a few minutes.

2.5 Interpretation of expression patterns

For each reporter gene fusion at least two independent transformed lines of *C. elegans* should be obtained prior to the interpretation of the expression pattern, to confirm that the expression pattern is not a consequence of a DNA rearrangement that occurred during *C. elegans* transformation. Expression of reporter gene fusions is often mosaic, and determination of an expression pattern requires the observation of many stained individuals. Part of this mosaicism may be attributed to exogenous DNA being transmitted in *C. elegans* as extrachromosomal arrays (see Chapter 5) with loss of the array during somatic cell divisions. However, integration of the array into a chromosome (see Chapter 5, *Protocol 7*) may not eliminate this complication.

Expression patterns, derived from the use of reporter genes, should only be considered a guide to the expression of the native gene. Reporter gene fusions may not contain all the regulatory elements that direct the expression pattern of the endogenous gene. Regulatory elements out of their natural context in the chromosome may not function properly. The fusion protein and the endogenous protein, or their corresponding transcripts, may have very different rates of turnover. Contrary to expectation, reporter gene expression in the germline and constitutive reporter-gene expression is rarely directed by fusion to *C. elegans* genes. (Recently, conditions more favourable for reporter gene expression in the germline have been described (14).) Cells of the posterior intestine and the pharynx show expression surprisingly often, perhaps as a result of the inappropriate expression of reporter genes (15, 16), perhaps when incomplete promoters are used in reporter gene fusions. Reporter gene fusions are useful for the rapid examination of gene expression patterns, but other techniques, such as those described below, are required to determine the definitive expression pattern of an endogenous gene.

3. Localization of RNAs in *C. elegans* embryos by whole mount, *in-situ* hybridization

The most direct method for determining the spatial and temporal expression of a given gene is to localize, by *in-situ* hybridization, the RNAs transcribed from this gene. Whole-mount, *in-situ* hybridization to embryos was first developed for *Drosophila* embryos (17). The procedure described here, adapted from other protocols then developed for *C. elegans* (18, 19), involves the use of single-stranded, digoxigenin-labelled RNA probes that offer several advantages including sensitivity, speed, safety, and single-cell resolution of the signal.

The procedure involves the following steps:

(a) synthesis of a labelled RNA probe;
(b) fixation and pre-treatment of the embryos;
(c) hybridization of the probe;
(d) washing to remove unbound probe;
(e) visualization of the hybridized probe.

Similar procedures have been developed for *in-situ* hybridization to the post-embryonic stages of *C. elegans* (20), but these are not described here.

3.1 Preparation of digoxigenin-labelled RNA probes

As described in *Protocol 4*, single-stranded RNA probes are synthesized by transcription of the required sequences cloned into the polylinker site of a plasmid vector, such as pBluescript and pCR-Script (Stratagen) or pGEM (Promega), that contain promoters for SP6, T7, or T3 bacteriophage RNA polymerases. The cDNA insert should be several hundred base pairs long. The vector is first linearized using an appropriate restriction enzyme. For an anti-sense probe, a unique restriction site immediately 5′ to the insert is used, and the digested DNA is transcribed using the promoter 3′ to the insert. For sense (control) probes, a unique restriction site immediately 3′ to the insert is used, and the digested DNA is transcribed using the promoter 5′ to the insert. Transcription is carried out in the presence of unlabelled ATP, UTP, GTP, CTP, and digoxigenin-UTP (Boehringer Mannheim). The amount of probe made (up to 10 μg from 1 μg of linear template DNA) can most easily be estimated by electrophoresing an aliquot of the transcription reaction product on an agarose gel containing ethidium bromide.

Protocol 4.   Synthesis of digoxigenin-labelled RNA probe

Equipment and reagents

- Agarose gel electrophoresis equipment
- Water bath at 37°C or 40°C
- 10 × transcription buffer: 400 mM Tris–HCl pH 8.25, 60 mM MgCl₂, 20 mM spermidine
- 0.2 M DTT
- Nucleotide mix: 10 mM ATP, 10 mM GTP, 10 mM CTP, 6.5 mM UTP, 3.5 mM digoxigenin-UTP (Boehringer Mannheim)
- 1 μg/μl linearized plasmid
- 100 U/μl ribonuclease inhibitor
- 20 U/μl RNase-free DNase I
- 10 U/μl SP6, T7, or T3 RNA polymerase
- TE buffer: 50 mM Tris–HCl, 1 mM EDTA, pH 8.0
- 3 M sodium acetate
- Ethidium bromide
- Ethanol
- 70% ethanol in water

Method

1. Mix the reagents in the following order at room temperature in an Eppendorf tube:

   - sterile distilled water   12.5 μl
   - 10 × transcription buffer   2.0 μl
   - 0.2 M DTT   1.0 μl
   - nucleotide mix   2.0 μl
   - linearized plasmid (1 μg/μl)   1.0 μl
   - ribonuclease inhibitor (100 U/μl)   0.5 μl
   - SP6, T3, or T7 RNA polymerase as appropriate (10 U/μl)   1.0 μl

2. Centrifuge briefly to collect the contents of the tube and incubate for 1 h at 37°C for T3 or T7 RNA polymerase, at 40°C for SP6 RNA polymerase.

3. Remove 1 μl and electrophorese, with a further 40 ng linearized plasmid in an adjacent lane, on a 1% agarose gel containing 0.5 μg/ml ethidium bromide. The yield of transcript can be estimated by comparing the relative staining intensity between the plasmid band and the RNA band.

4. Add 2 μl RNase-free DNase I (20 U/μl).

5. Incubate at 37°C for 15 min.

6. Add 100 μl TE buffer, 10 μl 3 M sodium acetate, and 300 μl ethanol, mix, and incubate at –20°C for 30 min.

7. Centrifuge at 11000 g (12000 r.p.m.) in a microcentrifuge at 4°C for 10 min, wash the pellet twice with 70% ethanol, and air-dry.

8. Redissolve the pellet in TE buffer at 100 ng/μl and store at –20°C.

3.2 Preparation of the embryos

Mixed-stage embryos are collected (see Chapter 4, *Protocol 5*) and attached on polylysine-coated slides (see *Protocol 2*). Embryos are then permeabilized by freezing and fixed as described in *Protocol 5*. Solutions and polylysine-coated slides should be prepared fresh on the day of use.

Protocol 5. Permeabilization and fixation of embryos

Equipment and reagents

- 8-well, multiwell microscope slides (ICN Biomedicals, cat. no. 6040805) coated with polylysine (see *Protocol 2*)
- Staining dishes
- 22 × 50 mm glass coverslips
- Aluminium block
- Dry ice
- 10 × PBS buffer: 80 g NaCl, 2 g KCl, 6.1 g Na$_2$HPO$_4$, 2 g KH$_2$PO$_4$·H$_2$O, per litre, pH 7.0
- PTw buffer: 1 × PBS, 0.1% Tween-20
- Methanol at –20°C
- Acetone at –20°C

Protocol 5. *Continued*

- 90% methanol in water
- 70% and 50% methanol in 1 × PBS
- 0.2 M HCl
- 1 μg/ml Proteinase K in PTw buffer

- 2 mg/ml glycine in PTw buffer
- Formaldehyde fixative solution: 1 × PBS, 0.08 M Hepes pH 6.9, 1.6 mM $MgSO_4$, 0.8 mM EGTA, 3.7% formaldehyde

Method

1. Transfer the embryos to the polylysine-coated slides (10 μl of embryos in water per well, use 4 wells per slide). Separate clumps of embryos on the slide with an eyelash attached to a toothpick, or blow air on to the embryos through a micropipette.

2. Cover the embryos with a glass coverslip, hanging the coverslip over the end of the slide for easier removal.

3. Freeze the embryos by placing the slides on an aluminium block that has been pre-cooled on dry ice. Leave for at least 10 min.

4. Flick off the coverslip and immediately immerse the slides in methanol at –20°C for 5 min.

5. Transfer the slides to acetone at –20°C for 5 min and then rehydrate at room temperature as follows (drain excess liquid between each step):
   - wash in 90% methanol in water, 1 min.
   - wash in 70% methanol in 1 × PBS, 1 min.
   - wash in 50% methanol in 1 × PBS, 1 min.
   - wash twice in Ptw buffer, 5 min each.

6. Transfer the slides to 0.2 M HCl for 20 min.

7. Wash twice in PTw buffer, 5 min each.

8. Transfer the slides to Proteinase K (1 μg/ml) for 15 min at room temperature.

9. Wash in glycine (2 mg/ml) for 2 min.

10. Wash twice in PTw buffer, 5 min each.

11. Immerse the slides in the formaldehyde fixative solution at room temperature for 20 min.

12. Wash at room temperature as follows:
    - wash twice in PTw buffer, 5 min each.
    - wash in glycine (2 mg/ml) for 5 min.
    - wash three times in PTw buffer, 5 min each.

3.3 Pre-hybridization and hybridization of the embryos

Embryos are pre-hybridized and then incubated overnight with a digoxigenin-labelled RNA probe (see *Protocol 6*). Following hybridization, slides are extensively washed to remove excess probe (see *Protocol 7*). To determine the probe concentration in the hybridization buffer that will yield the best signal-to-noise ratio, test serial dilutions of the probe—from 0.01 to 10 ng/μl.

Protocol 6. Pre-hybridization and hybridization

Equipment and reagents

- Slides from *Protocol 5*
- Humid incubator at 48°C
- Small paintbrush
- Vaseline
- Rubber O-rings
- Hybridization buffer I: 100 μg/ml autoclaved salmon sperm DNA, 50 μg/ml heparin, 0.1% Tween-20, 50% deionized formamide, 5 × SSC
- PTw buffer: 1 × PBS, 0.1% Tween-20

- 20 × SSC buffer: 175.3 g NaCl, 88.2 g Na₃ citrate·2H$_2$O, per litre, pH 7.0
- 1 × Denhardt's solution: 0.02% Ficoll 400, 0.02% polyvinyl pyrrolidone, 0.02% BSA
- Hybridization buffer II: 500 μg/ml autoclaved salmon sperm DNA, 1 × Denhardt's solution, 2 mM EDTA, 10% dextran sulfate, 50% deionized formamide, 4 × SSC
- Digoxigenin-labelled RNA probe (prepared using *Protocol 4*).

Method

1. Incubate the embryos for 10 min at room temperature in a 1:1 mix of hybridization buffer I and PTw buffer, and then for 10 min in undiluted hybridization buffer I. During this time, boil 10 ml of hybridization buffer II for 10 min and then cool on ice.

2. Carefully drain excess liquid from around the wells while keeping the embryos wet. With a small paintbrush put Vaseline around each well. Lay two rubber O-rings, the size of the wells, into the Vaseline.

3. Add 40 μl of freshly heated, hybridization buffer II per well and incubate for 1–2 h at 48°C in a humidity chamber. Meanwhile boil the probe diluted in hybridization buffer II for 1–2 h, to reduce the length of the probe, and then cool on ice.

4. Remove the pre-hybridization buffer and add 40 μl of the freshly heated diluted probe to each well and overlay with a coverslip coated with Vaseline[a].

5. Incubate the slides in a sealed humidity chamber overnight at 48°C.

[a] Alternatively cover each well with a square of Parafilm and then seal it on to the slide with rubber cement.

Protocol 7. Post-hybridization washes

Equipment and reagents

- Shaker
- Hybridization buffer I, as in *Protocol 6*, but without salmon sperm DNA: 50 μg/ml heparin, 0.1% Tween-20, 50% deionized formamide, 5 × SSC

- PTw buffer (see *Protocol 6*)
- 1 × PBS (see *Protocol 5* for 10 × PBS)
- PBT buffer: 1 × PBS, 0.1% BSA, 0.1% Triton X-100

Protocol 7. *Continued*

Method

Perform all washes with gentle agitation.

1. Remove the coverslip and rubber O-rings and place in a slide rack in hybridization buffer I at 48°C, for 30 min, replacing the buffer after 15 min.

2. Wash in 3 parts hybridization buffer I:2 parts PTw buffer at 48°C, twice for 15 min each.

3. Wash in 1 part hybridization buffer I:4 parts PTw buffer at 48°C, twice for 15 min each.

4. Wash in PTw buffer at 48°C, twice for 15 min each.

5. Wash in PBT buffer at room temperature, twice for 20 min each.

3.4 Immunochemical detection of the probe

After removal of the unbound probe, embryos are incubated with anti-digoxigenin antibody conjugated with alkaline phosphatase. After washing, detection of the alkaline phosphatase (see *Protocol 8*) with a highly sensitive, colour reaction generates stained specimens that can be stored permanently (see *Figure 3*). Fluorescent detection has also been used to determine the subcellular localization of relatively abundant RNAs (18).

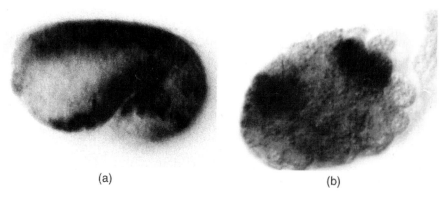

(a)   (b)

Figure 3. Localization of *unc-54* and *pes-1* mRNA in wild-type N2 *C. elegans* embryos by whole-mount *in-situ* hybridization as in *Protocols 4–8*. Embryos were hybridized to an *unc-54* or *pes-1* RNA probe and visualized using alkaline-phosphatase-mediated detection (dark in the figure). (a) The *unc-54* probe reveals the distribution of body-wall myosin transcripts in differentiating muscle cells of an elongation-stage embryo. (This panel was kindly provided by G. Sowa and R. Schnabel.) (b) The *pes-1* probe gave much weaker staining than the *unc-54* probe and in younger embryos.

Protocol 8. Alkaline phosphatase-mediated probe detection

Equipment and reagents

- Alkaline phosphatase-conjugated, anti-digoxigenin antibody (Boehringer Mannheim)
- Staining solution: 100 mM NaCl, 5 mM MgCl$_2$, 100 mM Tris pH 9.5, 0.1% Tween-20, 1 mM levamisole[a]
- NBT (4-nitro blue tetrazolium chloride; Boehringer Mannheim)
- DAPI (diamidinophenolindole) (Sigma)
- X-phosphate (5-bromo-4-chloro-3-indolyl-phosphate; Boehringer Mannheim)
- 1 × PBS (see *Protocol 5* for 10 × PBS)
- PBT buffer (see *Protocol 7*)
- Mounting medium: 90% glycerol, 1 × PBS; add 1% *n*-propyl gallate if DAPI is used

Method

1. Apply 30 μl of the diluted alkaline phosphatase, anti-digoxigenin antibody conjugate (1:1500 in PBT) to each well and incubate in a humidity chamber for 2 h at room temperature or overnight at 4°C.

2. Wash 4 times for 10 min each in PBT.

3. Incubate the embryos for 10 min in freshly made staining solution without NBT and X-phosphate.

4. Apply 15 μl of the staining solution containing 4.5 μl/ml of NBT, 3.5 μl/ml of X-phosphate, and 1 μg/ml of DAPI to each well. Incubate the slides at room temperature in the dark for between 20 min and 24 h, depending on the probe.

5. Stop the colour reaction by washing the embryos in 1 × PBS for 10 min at room temperature.

6. Apply 10 μl of mounting medium to the embryos and cover with a glass coverslip.

[a] Levamisole is included to inhibit endogenous alkaline phosphatases.

---

4. Immunomicroscopy

While *in-situ* mRNA hybridization can reveal when and where a gene is expressed, the active product of a gene is usually the protein encoded by that mRNA. The protein may be much more stable than the mRNA and remain active long after the gene is no longer expressed. The protein may be localized to particular sites within the cell. The protein may even be transported out of the cell to act at a site distant from the point of synthesis. Therefore, the distribution of the protein product is likely to be more important for an understanding of the biological role of a gene than knowledge of when and where that gene is expressed. Immunomicroscopy, using antibodies to determine the location of a protein, can provide this information. A specimen is

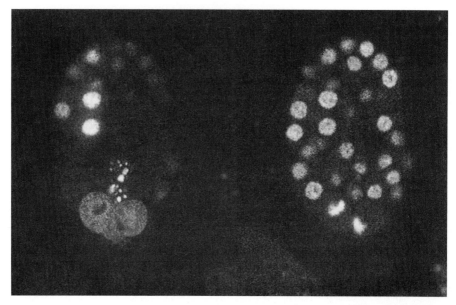

Figure 4. Confocal images of a 100-cell, UL1, *C. elegans* embryo (23). UL1 was generated by transformation of the wild-type strain N2 with a *pes-1::lacZ* fusion gene. The preparation was stained as in *Protocol 9* with two monoclonal antibodies, one recognizing β-galactosidase and one recognizing the P-granules. In the left panel (green fluorescence), the tight foci of staining towards the centre of the embryo are the P-granules that segregate to the germline during embryonic cell divisions and are in the two cells Z2 and Z3 at this stage. The β-galactosidase is nuclear localized in the cells of the AB cell lineage at the top of the panel, but is spread throughout the cells Da and Dp at the base of the panel because these cells are about to divide and the nuclear envelope has already broken down. In the right panel (red fluorescence), all the nuclei are visible through staining of the DNA with propidium iodide. The condensed chromosomes of the cells Da and Dp exclude the β-galactosidase.

stained with an antibody preparation that specifically recognizes the protein of interest, and the distribution of that antibody is then visualized (see *Figure 4*).

Starting from the cloned and sequenced gene, as is now possible for all *C. elegans* genes, there are many standard procedures for expressing the protein in a recombinant system. This can then be used as an antigen to raise antibodies specific to that protein (21). However, obtaining antibodies that recognize *C. elegans* proteins has been found to be difficult. Furthermore, background problems, with an antibody preparation recognizing other *C. elegans* antigens non-specifically, have often been encountered. Cross-reactivity to related *C. elegans* proteins may also be observed. Specificity of a staining reaction can be confirmed by signal loss in a control specimen derived from a strain bearing a mutation in the gene of interest. Procedures for generating the immunological reagent will not be described here.

Each antibody–antigen interaction is unique and may be more favoured by certain conditions. *Protocol 9* (based on ref. 22) describes one procedure that is generally appropriate for *C. elegans* preparations. Alternative fixation procedures (e.g. 3% formaldehyde or heating to 100°C for 2 min) may make an antigen more accessible or better recognized by some antibodies. The dilution, temperature, and buffer conditions at step 6 may need to be optimized for any particular antibody–antigen interaction (e.g. incubate at room temperature or 4°C instead of 37°C, or the time may be increased to 24 h). An appropriate procedure for any new antibody can only be determined by trial and error.

Protocol 9. Indirect immunofluorescence microscopy

Equipment and reagents

- Worms or eggs prepared as described in Chapter 4
- M9 buffer (see *Protocol 1*)
- 8-well, multiwell microscope slides (ICN Biomedicals, cat. no. 6040805). Sub by coating with polylysine (see *Protocol 2*) or by dipping in 0.05 mg/ml BSA and dried.
- Coverslips
- Methanol and acetone in Coplin jars at –20°C
- Primary antibody specific to the protein of interest

- Fluorescein-conjugated secondary antibody that will recognize the primary antibody (Sigma provide a range of secondary antibodies)
- PBS: 0.15 M NaCl, 10 mM phosphate, pH 7.2
- PBSA: 1 × PBS, 3% BSA
- 10 mg/ml RNase A
- Mounting medium: 90% glycerol, 0.1 × PBS, 25 mg/ml DABCO (Sigma), 25 μg/ml propidium iodide (Sigma)
- Epifluorescence microscope and appropriate filter sets

Method

1. Apply the eggs or worms to the subbed microscope slide in M9 buffer, 3 μl per well.

2. Apply a coverslip. Squash adult worms slightly or they will not adhere to the subbed slides.

3. Freeze rapidly on dry ice, then flip off the coverslip and place the slide immediately into methanol at –20°C, for 5 min.

4. Transfer the slide to acetone at –20°C for 5 min.

5. Place the slide briefly in PBS.

6. Dry around the wells with a tissue and apply the antibody diluted[a] in PBSA, 10 μl per well. Incubate at 37°C, for 2 h, in a humidity chamber.

7. Place the slide in PBS for 30 min, with three changes of the buffer.

8. Dry around the wells with a tissue and apply the secondary antibody diluted[a] in PBSA containing 12.5 μg/ml RNase A, 10 μl per well. Incubate at room temperature for 2 h in a humidity chamber.

9. Place the slide in PBS for 30 min, with three changes of the buffer.

10. Mount the specimen in the mounting medium.

Protocol 9. *Continued*

11. Observe the specimen under epifluorescence using filter sets appropriate for visualizing the green emission of the fluorescein (antibody/specific protein) (e.g. Zeiss filter set 9) or the red emission of the propidium iodide (DNA) (e.g. Zeiss filter set 15).

[a] The primary antibody may need to be diluted only 2-fold if it is a hybridoma supernatant or 1000 × if it is a concentrated monoclonal antibody preparation. Too high a concentration will give a high background. Too low a concentration will give a weak signal. The secondary antibody is normally diluted 50–100 × depending on the manufacturer's recommendation.

5. Conclusions

Patterns of gene expression provide a direct link between the genome sequence data and the developmental anatomy of *C. elegans* as observed with a microscope. This direct connection and the ease with which such data can be generated may mean gene-expression pattern information will have a crucial role in our bid to interpret the genome sequence data. Nevertheless, even the protein distribution, as revealed by immunomicroscopy, does not necessarily reveal where and when a gene's ultimate activity is expressed. A protein may be locally activated post-translationally, or the distribution of the target of a protein's activity may not match the distribution of the protein (e.g. proteins with signalling roles in the control of development). Inferences about gene function based purely on gene-expression pattern data must be cautious. Determination of genetic function requires the integration of results obtained using the other approaches outlined in other chapters in this book.

References

1. Fire, A., Harrison, S. W., and Dixon, D. (1990). *Gene*, **93**, 189.
2. Silhavy, T. J. and Beckwith, J. R. (1985). *Microbiol. Rev.*, **49**, 398.
3. Prasher, D. C., Eckenrode, V. K., Ward, W. W., Prendergast, F. G., and Cormier, M. J. (1992). *Gene*, **111**, 229.
4. Jefferson, R. A., Klass, M., Wolf, N., and Hirsh, D. (1987). *J. Mol. Biol.*, **193**, 41.
5. Goodwin, E. B., Okkema, P. G., Evans, T. C., and Kimble J. (1993). *Cell*, **75**, 329.
6. Roberts, B. L., Richardson, W. D., and Smith, A. E. (1987). *Cell*, **50**, 465.
7. Brinster, R. L., Allen, J. M., Behringer, R. R., Gelinas, R. E., and Palmiter, R. D. (1988). *Proc. Natl Acad. Sci. USA*, **85**, 846.
8. Buchman, A. R. and Berg, P. (1988). *Mol. Cell. Biol.*, **8**, 4395.
9. Fire, A. (1992). *Genet. Anal. Tech. Appl.*, **9**, 152.
10. Chalfie, M., Tu, Y., Euskirchen, G., Ward, W. W., and Prasher, D. C. (1994). *Science*, **263**, 802.
11. Cubitt, A. B., Heim, R., Adams, S. R., Boyd, A. E., Gross, L. A., and Tsien, R. Y. (1995). *Trends Biochem. Sci.*, **20**, 448.

12. Heim, R., Prasher, D. C., and Tsien, R. Y. (1994). *Proc. Natl Acad. Sci. USA*, **91**, 12501.
13. Heim, R., Cubitt, A. B., and Tsien, R. Y. (1995). *Nature*, **373**, 663.
14. Kelly, W. G., Xu, S., Montgomery, M. K., and Fire, A. (1997). *Genetics*, **146**, 227.
15. Hope, I. A. (1991). *Development*, **113**, 399.
16. Krause, M., White Harrison, S., Xu, S., Chen, L., and Fire, A. (1994). *Dev. Biol.*, **166**, 133.
17. Tautz, D. and Pfeifle, C. (1989). *Chromosoma*, **98**, 81.
18. Seydoux, G. and Fire, A. (1995). In C. elegans: *modern biological analysis of an organism* (ed. H. Epstein and D. Shakes), p. 323. Academic Press, San Diego, CA.
19. Tabara, H., Motohashi, T., and Kohara, J. (1996). *Nucleic Acids Res.*, **22**, 2119.
20. Harlow, E. and Lane, D. (1988). *Antibodies: a laboratory manual.* Cold Spring Harbor Laboratory Press, NY.
21. Birchall, P. S., Fishpool, R. M., and Albertson, D. G. (1995). *Nature Genet.*, **11**, 314.
22. Sulston, J. and Hodgkin, J. (1988). In *The nematode* Caenorhabditis elegans (ed. W. B. Wood), p. 599. Cold Spring Harbor Laboratory Press, NY.
23. Hope, I. A. (1994). *Development*, **120**, 505.

10

Molecular biology

IAIN L. JOHNSTONE

1. Introduction

This chapter deals with some of the common, molecular biological methods essential for general aspects of gene characterization. It includes the purification of DNA, RNA, and some simple, protein extraction methods useful for Western blot analyses. Because much of what is discussed here is not specific to the nematode and extensive coverage is available in other texts (1), I have emphasized the modifications to those more general methods relevant to working with the worm.

The availability of the genome sequence makes an immense body of data available to any researcher. It is an extremely valuable resource which, for many of us in the field, has resulted in a great reduction in the amount of general molecular biology performed in the laboratory. However, it is important to avoid complacency and remember that most of the genes indicated by the genome project are predictions, and that the different aspects of such predictions, e.g. intron/exon organization and gene endpoints, must be confirmed experimentally. This is particularly important when considering the construction of reporter gene fusions and the interpretation of expression patterns so obtained (see Chapter 9). The predictions, around the start of genes in particular, are not always accurate. In this post-genome sequence era, in addition to confirming predicted data, general molecular methods are still necessary for many purposes including the analysis of gene expression and gene function, as well as for investigating the nature of mutant strains.

2. The raw material—worms

The starting point for the methods discussed here is healthy, living worms. Culture methods are detailed in Chapter 4. We grow C. elegans by feeding them bacteria, either on lawns on agar plates or in liquid culture. Liquid culture is useful for the production of larger numbers of worms, but, in practical terms, it can be rather awkward. For some purposes, the bacteria used to culture worms can be grown in a standard bacterial medium, collected by centrifugation, and added in concentrated form to *C. elegans* plate cultures. This permits the generation of much greater masses of worms from plates than is possible by standard plate culture methods, but there is an upper limit where toxic by-products and/or an anaerobic environment start to kill the worms. The addition of 2 ml of a bacterial slurry to a standard healthy culture growing on a 10 cm plate can increase the number of animals that can be obtained, but the worms should be monitored under a dissecting microscope to ensure they stay healthy.

In general, worms can be collected from liquid culture by centrifugation, or from plate culture by first washing them off in M9 buffer followed by centrifugation (see Chapter 4 *Protocol 3*). Worm pellets are not very robust and the movement of the animals will cause the pellets to disperse quite rapidly. This problem can be minimized by chilling the buffers used.

The use of bacteria as worm food means that worms collected by centrifugation are invariably contaminated by bacteria and bacterial biomolecules, which can be a problem for some purposes. The two sources of potential bacterial contamination are bacteria outside the worms (either in the liquid culture medium or washed off the agar plates along with cultured worms), and bacteria inside the worm's gut. Some ways of reducing this contamination are discussed below.

Finally, if the worms are to be used for DNA preparations, do not use agar in plate culture. The worms should either be grown in liquid culture or agarose should be used in place of agar on plates (simply substitute agarose in the medium in Chapter 4 *Protocol 2*). Agar can cause problems when DNA is to be subsequently digested with restriction enzymes.

2.1 External contamination

External contamination can be greatly reduced by either filtering the worms as in *Protocol 1* or using the sucrose flotation method described in *Protocol 2*. For many purposes, contamination with some bacteria is not a problem, so long as the presence of such material is considered when designing experiments and interpreting results. Collecting worms from just-starved cultures can be a simple and convenient way of limiting bacterial contamination.

2.2 Internal contamination

Bacterial material in the guts of worms is removed by allowing completion of digestion without further ingestion. Worms that have been externally cleaned by one of the two protocols below can be incubated in M9 buffer at 20°C for an hour, to permit digestion of material in the gut. This should be carried out in a suitable container to permit a high surface-to-volume ratio of the M9 buffer. The worms are prone to becoming oxygen starved and so should be checked frequently to monitor their movement. Worms can be collected by centrifugation or further filtration, but this procedure could influence gene expression.

2.3 Factors influencing gene expression

Where worms are to be used for gene expression studies (RNA or protein extraction), it is essential to be familiar with worm developmental biology and to consider the effect that any worm culture or purification procedures may have on worm gene expression.

The worm life-cycle includes an embryonic, four larval, and adult stages (see Chapter 1). The proportion of the various developmental stages present in the starting material can vary greatly. Obviously, this can be very important if, for example, RNA is to be prepared to examine gene expression. Cultures can become rather synchronous if allowed to starve, or if they are started using an inoculum of a particular developmental stage. (This can be put to use as discussed later.) Larvae and adults are easily washed off plate cultures with M9 buffer or water, however embryos have a tendency to stick to the agar surface. There are a variety of ways in which the relative proportion of developmental stages in the starting material can vary.

The treatment given to worms during collection, before RNA or protein isolation, must also be considered. Although collecting worms from just-starved cultures reduces contaminating bacteria, in addition to causing the population to be non-representative for the various developmental stages, starvation itself will influence gene expression. This could also be true if efforts are taken to remove the internal contamination of bacteria as discussed in Section 2.2 above. All methods of purifying worms from contaminating bacteria may influence gene expression studies. The sucrose flotation method (see *Protocol 2*) is very effective in providing clean worms; however, the worms are stressed by this procedure. This method is useful where the material is to be used for DNA extraction, but is not advised where gene expression is important. The filtration methods of collection and washing are preferable where gene expression is important. However, even with this method there can be some differential loss of certain developmental stages. Any mesh used for filtration should be tested to see if smaller larvae tend to escape.

In some cases, it may be best to accept a degree of bacterial contamination. If working with small numbers of worms, bacterial RNA can act as a good carrier in nucleic acid purification and worm mRNA can be readily purified from the mixture by the selection of poly-A tailed RNA. (Poly-A tails are present on eukaryotic but not prokaryotic mRNA.)

Protocol 1. Filtration of worms with nylon mesh

Equipment and reagents
- Filter mesh[a] (Cadisch Precision Meshes Ltd)

Protocol 1. *Continued*

A. *Large scale*

1. Cut a suitably sized piece of nylon mesh and place in a clean plastic funnel.

2. Filter the worms through the nylon mesh.

3. Wash the worms with several volumes of water to remove the culture medium and bacteria.

4. Collect the worms either by scraping them directly off the surface of the mesh with a spatula into suitably sized centrifuge tubes, or washing off the mesh into a beaker with chilled water[b] from a squeezy bottle (preferably kept on ice prior to use).

5. If they are washed off, concentrate the worms by centrifuging at 500 *g* in a benchtop centrifuge for 5 min or for 30 sec in a micro-centrifuge.

6. Remove the supernatant immediately by pipette. Do not pour off the supernatant as worm pellets are not robust.

B. *Small scale*

Filtering small volumes using the above method may result in the loss of too many worms, but a variety of small filters can be constructed using the same mesh.

1. Cut a hole in the centre of the lid of a microcentrifuge tube, large enough to allow liquid to drain through, and slice off the narrowed region towards the bottom of the tube. (Use a scalpel blade heated in the flame of a Bunsen burner.)

2. Close the lid over a small piece of mesh placed between the body of the tube and the lid, thus trapping the mesh in place. This makes an effective disposable worm filter.

3. Filter the worms by inverting the tube thus prepared, pipetting the worms into the upturned and opened bottom of the tube, and allowing liquid to pass through the mesh and exit through the hole cut in the lid. The process can be speeded up by applying positive or negative pressure.[c]

---

[a] Nylon 10 micron is effective at retaining all developmental stages—however, the flow-through is rather slow. Alternatively, 20-micron or 50-micron nylon can be used if some loss of smaller larvae is not a problem. Similar material can be sourced from companies that supply materials for screen printing.
[b] Chilling the water is useful as it stops the worms from moving and disrupting the pellet.
[c] We have also made useful worm filters by cutting the end off a disposable syringe barrel and gluing a circle of mesh to the cut barrel using a standard epoxy resin.

Protocol 2. Sucrose flotation

Equipment and reagents
- 0.1 M NaCl
- 60% (w/v) sucrose (ice-cold)

Method

1. Collect worms by centrifugation (see Chapter 4 *Protocol 3*).

2. Resuspend the pellet in 10 volumes of 0.1 M NaCl.

3. Add an equal volume of the cold 60% sucrose solution and mix by inversion.

4. Centrifuge immediately at 500 *g* for 2 min in a pre-cooled benchtop centrifuge.

5. Remove the floating cap of worms with a Pasteur pipette cut at the shoulder.

6. Immediately dilute the sucrose in the collected worms by adding 10 vol. of cold 0.1 M NaCl.

7. Centrifuge as in step 4.

8. Remove the supernatant.

9. Resuspend the worm pellet in 50 vol. of 0.1 M NaCl at 20°C.

10. Place the worms in a shallow dish for good aeration. Examine with a dissecting microscope to check that debris is removed and the worms are moving well.

11. Incubate for 1 h with agitation to encourage the digestion of bacteria in the worm guts.

12. Collect worms by centrifuging at 500 *g* for 2 min in a pre-cooled bench centrifuge.

---

3. DNA methods

3.1 General considerations

DNA can be prepared from the worm with very minor modifications of the methods used for the preparation of DNA from whole tissues from other organisms (1). Worms are contained within a tough exoskeleton that tends to resist mechanical shearing. Some methods that make use of certain mechanical homogenizers fail because the animals do not break open. Grinding in liquid nitrogen (see *Protocol 3*) is a reliable method for fragmenting animals and providing material suitable for a variety of extraction methods. Alternatively, the worms can be subjected directly to proteolytic digestion and reduction, which solubilizes the exoskeleton and other proteins. Higher molecular weight DNA is obtained by methods that start with grinding in liquid nitrogen. The direct digestion methods are convenient if many samples have to be prepared, and are certainly adequate for most purposes including Southern blotting.

The influence of bacterial DNA contamination should also be considered, as discussed above. The developmental stage is not crucial for DNA preparation, but the expansion in the germline nuclei in L4 and adult animals may improve yields. The worms should be alive and healthy immediately before the preparation begins, otherwise the prepared DNA may be degraded.

Protocol 3. DNA isolation

Equipment and reagents

- Pestle and mortar
- Liquid nitrogen, in an appropriate container suitable for pouring small amounts
- Protective glasses and gloves
- DNA lysis buffer: 0.1 M NaCl, 10 mM Tris–HCl pH 8.0, 10 mM EDTA, 1% SDS, 1% β-mercaptoethanol, 100 μg/ml Proteinase K
- 50 ml Falcon tubes
- Chloroform

- Phenol[a] buffered with Tris pH 8.0 (and preferably containing some 8-hydroxyquinoline)
- 3 M sodium acetate pH 4.5
- Isopropanol
- 70% ethanol
- TE buffer: 1 mM Tris–HCl pH 8.0, 1 mM EDTA

A. *Large-scale grinding in liquid nitrogen*

Caution: Liquid nitrogen is a considerable safety hazard. Protective glasses should always be worn.

1. Collect worms by the sucrose flotation method (see *Protocol 2*) then incubate to digest bacteria in the gut.[b]
2. Concentrate the worms into a slurry, either by centrifugation or filtration as in *Protocol 2*. Estimate the volume of worms at this stage and use this later to calculate the appropriate volume of lysis buffer needed (see step 5).
3. Place a mortar and pestle on top of a sheet of aluminium foil (to catch frozen lumps of worms which are jettisoned from the mortar) and cool by filling the mortar with liquid nitrogen. Refill with nitrogen as necessary.
4. Add the worms dropwise to the liquid nitrogen in the mortar either using a pipette or by scraping with a spatula. Aim to freeze the worms in small distinct droplets. Make sure they do not thaw subsequently, at any point during this process.
5. Measure at least 10 volumes of the lysis buffer (with reference to the 1 volume of worms, as measured in step 2 above), into a suitably wide container (e.g. a 50 ml Falcon tube). Pre-heat the lysis buffer to 60°C. Use in step 7.

6. Use the pestle to gradually break up the beads of worms under nitrogen. Grind more forcibly into a fine powder when the beads are considerably fragmented and when these fragments tend to stay in the mortar during grinding. Let the nitrogen almost evaporate at this stage so that the worms take on a paste-like texture with the nitrogen. (They grind well in this condition.) Keep adding nitrogen as necessary. Grind for 2 to 3 min.

7. Allow the nitrogen to evaporate. Disperse the deeply frozen powder rapidly in the DNA lysis buffer by scraping the powdered worms on to the surface of the pre-heated buffer with a clean spatula, pre-cooled in liquid nitrogen. Make sure that the ground worms are not allowed to warm up prior to their rapid dispersal in lysis buffer. Avoid trans-ferring residual liquid nitrogen at this stage as it boils instantly on contact with the buffer and is a considerable safety hazard. **Always wear protective glasses.** Transfer the frozen powder just as the nitrogen boils off and use a clean glass rod or disposable plastic pipette to promote the dispersal of the powder. Take particular care if very high molecular weight DNA is required.

8. Once dispersed, transfer the mixture to a suitable tube, then cap, and incubate it at 60 °C for 1 h, with occasional gentle mixing.

At this stage, a variety of methods can be used to purify the DNA. A standard organic extraction method is given below.

B. *Direct digestion of worms in lysis buffer*[c]

1. Avoid the liquid nitrogen step by adding live intact worms directly to the correct volume of lysis buffer (pre-warmed to 60 °C) as calculated in Part A, step 2.

2. Follow Part A, step 8.

C. *Organic extraction purification of DNA*[d, f]

1. Extract the lysed worm preparation with 1 volume of phenol. Separate the phases by centrifuging at 500 *g* for 5 min, collecting the upper aqueous phase. Back-extract the interphase with a small volume of fresh lysis buffer, as a very high percentage of the DNA sometimes associates with the interphase.

2. Extract the aqueous phase twice with phenol/chloroform being careful to avoid the white material at the interface.

3. Extract the aqueous phase once with chloroform.

4. Add 0.1 volume of 3 M sodium acetate pH 4.5 and 0.7 volumes of isopropanol, mix gently by inversion, and allow the DNA to precipitate at room temperature for 5 min (precipitation usually occurs im-mediately).

Protocol 3. *Continued*

5. Collect the DNA by centrifuging at 500 *g* for 5 min, then wash the pellet with 70% ethanol.

6. Resuspend the nucleic acid in TE buffer and leave at room temperature for at least 1 h to allow dissolution.

7. **Optional.** Add DNase-free RNase A at a final concentration of 10 μg/ml and incubate at 37°C for 1 h if the removal of RNA is important[e] Then perform a second precipitation with isopropanol (after the addition of 0.1 volume sodium acetate) from step 4.

---

[a] Only high-quality, molecular biology grade phenol should be used.

[b] This is essential to avoid contamination with DNA from other organisms.

[c] The intact worms dissolve quite rapidly during the 60°C incubation. The molecular weight of the DNA obtained in this way is generally high enough for most purposes such as Southern blotting and PCR; but where very high molecular weight is important, grinding in liquid nitrogen is more reliable.

[d] Organic extractions are described in some detail in ref. 1.

[e] There is considerable contamination with RNA at this stage; this step also assists in dissolution.

[f] *Alternative purification methods*: It is possible to make use of a variety of commercial DNA purification kits and it should be possible to adapt the use of most kits designed for use with animal tissues for use with the worm. The major problem in obtaining high molecular weight DNA from *C. elegans* is simply the speed with which the worms are dissolved. The Qiagen Genomic-tips system can be adapted for use. Grind the worms first in liquid nitrogen as described above then treat according to the manufacturer's methods for tissue. The frozen ground worms are rapidly dispersed in the protease-containing lysis buffer described in the protocol booklet provided by the manufacturer. Proteinase K is suitable, but must be purchased separately. (Its use is described in the protocol, but is not provided with the columns).

Purified DNA should be electrophoresed on agarose gels using standard methods (1) to estimate quality and purity with respect to contaminating RNA (the smear that runs at the bottom of the gel if the RNase A step is omitted). A comparison with standards run on the same gel can be used to estimate concentration and size. If the sample is RNA-free, the standard OD_{260} method for estimating the DNA concentration (1) can be used. The DNA should be of sufficient quality for restriction digestion, Southern blot analysis, PCR, etc.

3.2 Southern analysis

Any general molecular biology source should be consulted for restriction digestion of genomic DNA, electrophoresis methods, and blotting methods (1). For Southern blots, we routinely use Hybond-N membrane (Amersham) and follow the methods described by the manufacturer for the treatment of gels and blotting. Any standard method will work. Where transfer is to be performed by the original capillary method, treating the gel with HCl improves transfer.

For most blotting purposes, digest approximately 1 μg of *C. elegans* genomic DNA with an appropriate restriction enzyme and subject the fragments to electrophoresis. This is sufficient DNA to allow the easy detection of single-copy sequences by any standard detection system. The concentration of agarose gel used for electrophoresis will depend on the molecular weight of the expected DNA fragments, but 0.8% is a good starting point. The gel should be stained to visualize DNA before blotting to confirm that the digestions have worked. With some enzymes, faint bands of repetitive DNA can be seen superimposed on top of the general smear of restricted genomic DNA and this is generally a good indicator of complete digestion of the sample.

3.3 Genomic PCR analysis

Standard texts should be consulted for general PCR methods, including the detection and cloning of sequences from genomic DNA (2). We routinely use approximately 0.1 μg of genomic DNA for PCR reactions where the intention is to clone the amplified product; however, it is possible to use very much less. Any of the above methods will give DNA well suited for PCR. PCR can also be performed very successfully on less pure samples of DNA, including simple, single animal lysates (see Chapter 6).

4. RNA methods

4.1 General considerations on worm gene expression

With the completion of the genome project, the genomic sequence of any gene will invariably be available. *C. elegans* cosmid sequences available in databases such as EMBL or in ACeDB contain, unless otherwise stated, *predictions* of genes and gene structure. These predictions must be confirmed experimentally, particularly with respect to start and end points and intron/exon structure, and this will involve transcript analysis. When starting any research on the expression of a worm gene, a degree of familiarity with worm biology is necessary. Consideration should be made of the numbers and different types of cells present in the animal. It is only possible to make whole animal RNA extracts, and hence the transcript of a gene that is only expressed in, for example, a few touch receptor cells may be present at very low abundance in RNA from the whole animal. Consideration should be given to the discussion in Section 2.3 above on the influences on gene expression by various worm purification procedures and the relative abundance of the different developmental stages in any culture used for RNA extraction. For some purposes, the generation of synchronous cultures provides a very convenient means of obtaining RNA that is representative of a narrow window of developmental time. This can be very useful where stage-specificity of gene expression is to be investigated, and it is discussed in more detail later.

However, for many basic gene characterization purposes, RNA that is representative of all normal development is useful. For this purpose, a truly non-synchronous, mixed-stage culture—containing all stages from embryos through to egg-laying adults—is the necessary starting material. In all cases, the simplest and least stressful (to the worms) method of harvesting should be used. Finally, there are two very significant developmental forms that are basically absent from typical wild-type *C. elegans* cultures, dauer larvae and males. If you cannot find a transcript corresponding to your chosen gene in mixed-stage RNA, think of looking in one of these two developmental routes.

4.1.1 The dauer larva

In response to starvation in the L1 and L2 stages, *C. elegans* can form a resistant alternative L3-stage larva termed the dauer (3). Cultures that have been left to grow too long will be almost free of bacteria and will contain large old adults, very few eggs, and an abundance of small and thin looking larvae. These larvae are a mixture of pre-dauer L1 and L2 larvae, and dauer L3s. If the culture is left longer, only L3 dauer larvae will remain. For most gene expression studies of the dauer developmental pathway, it is probably desirable to have both the L3 dauer larvae, and a mixture of the L1 and L2 larvae *en route* to forming dauers. A just-starved culture is best for this. RNA can be prepared from such a culture by the same method as given below for mixed-stage cultures.

4.1.2 The other sex

C. elegans has two sexes, the self-fertilizing hermaphrodite and the male (3, 5). Unless steps are taken to intentionally generate cultures containing males, wild-type cultures will typically be less than 1% male. When considering the transcription of a gene, it may be unwise to ignore the significantly different, male developmental pathway. There are male-specific cells, including neurons, male-specific structures, and male-specific patterns of behaviour. Male development is an obvious place to check for a transcript that cannot be detected in a normal mixed-stage culture. Whether it is safe to assume that a transcript found in the hermaphrodite represents the only transcript of a gene is a decision that must be made by the individual researcher based on the biology of that particular gene. However, I am unaware of any documented cases of sex-specific alternative transcripts/splicing variants for any gene in *C. elegans*.

There are two common ways to generate cultures with significant numbers of males. The simplest is to use a *him-8* mutant strain in which hermaphrodites produce approximately 37% male progeny (6). These can be cultured either on plates or in liquid culture. The *him-8* males are true males, the result of X chromosome non-disjunction in their mother, so they are perfectly valid for studies on gene expression in males. Males can also be generated from wild-type strains (see Chapter 12). If males and hermaphrodites are cultured in liquid, no mating will occur and the subsequent generation will be almost devoid of males, the result of 'selfing' by the hermaphrodites. For this reason, cultures of wild-type males must be grown on plates. Liquid culture can be used for *him-8* strains as the generation of males in this background does not involve mating. These methods generate mixed male/hermaphrodite cultures. There are some behavioural assays that can be used to partially purify males (3, 7).

4.1.3 Worm introns, operons, and *trans*-splicing

For those not familiar with worm molecular genetics, the organization of transcription units in the worm requires some mention here. A good review of gene structure and RNA processing is given in ref. 4. Many *C. elegans* genes are, like those of most other eukaryotes, organized in single-gene transcription units and frequently contain introns that are *cis*-spliced during the formation of the mature transcript. *C. elegans* introns are typically small, often 40–100 bases long; however, larger introns do occur. There are examples of differential splicing, the possibility of which should be considered when characterizing any gene transcript. As with other eukaryotes, the mature mRNA of worm genes is poly-adenylated and can be purified from total RNA by standard poly-A selection methods.

In addition to normal *cis*-splicing, *C. elegans* also has a process described as *trans*-splicing, in which a short RNA species, the spliced leader (SL), is spliced on to the 5' end of some mRNAs. Approximately 70% of all *C. elegans* genes are *trans*-spliced. This process has been reviewed elsewhere (4, 8). The most common *trans*-spliced leader is the 22 nucleotide SL1. There are two obvious consequences of this process: the 5' end of the mRNA in approximately 70% of all *C. elegans* genes is not co-incident with the transcriptional start sites; and the mRNA of a large proportion of all *C. elegans* genes share a common 22 nucleotide sequence at their 5' ends. Special consideration must therefore be made when attempting to determine the transcriptional start sites of worm genes. The sequence of the two most common *trans*-spliced leaders in *C. elegans* is given below. SL1 is the most common. In addition to the SL2 sequence given below, there are also some rare variants of this.

```
SL1 GGTTTAATTACCCAAGTTTGAG
SL2 GGTTTTAACCCAGTTACTCAAG
```

A second unusual phenomenon found in nematodes, related to *trans*-splicing, is the existence of operons (4, 8). Some worm genes exist in clusters and may be co-expressed from a single primary transcript. (In some cases, the existence of smaller primary transcripts containing only the downstream gene cannot be excluded.) Although probably the majority of *C. elegans* genes exist as mono-cistronic transcription units, the existence of these worm operons must be considered in relation to the transcriptional analysis of worm genes. For those cases known, the intergenic region between a downstream gene and its upstream partner is usually between 80 and 400 bases. The downstream gene is always *trans*-spliced. The *trans*-splicing process appears to be part of the process whereby mature translatable mRNA is generated for such genes. The downstream genes of *C. elegans* operons are most commonly *trans*-spliced, either exclusively to the alternate *trans*-spliced leader termed SL2, or to a mixture of both SL1 and SL2. SL2 *trans*-splicing is believed to be exclusive to the downstream genes of poly-cistronic mRNAs. The SL2 *trans*-spliced leader is not found on genes that are mono-cistronic or on genes that are the most upstream of a poly-cistron (4, 8). Where any predicted gene has a close upstream neighbour, the possibility of a single upstream promoter driving the expression of both genes, should be considered.

For practical purposes, the sequence of the *trans*-spliced leaders can be used in oligonucleotides for the PCR-based amplification of the 5′ ends of those genes that are *trans*-spliced. This is a modified 5′ RACE (rapid amplification of cDNA ends) reaction, detailed in Section 4.5 below.

4.2 RNA purification methods

As with RNA isolation from other organisms, the most intact RNA is obtained with methods that permit a very rapid dispersal of cellular material in a solution that inactivates cellular RNases. For worms, the eggshell of embryos and the cuticle of larvae and adults must be permeabilized or solubilized to permit the extraction of RNA and the inactivation of RNases. A very effective method is to grind the worms in liquid nitrogen, then rapidly disperse this material in a strongly denaturing RNA isolation buffer. Many commercial reagents/kits can be used in this way, as can some standard RNA isolation methods. This is discussed in *Protocol 4* below.

Protocol 4. Large-scale RNA isolation[a]

Equipment and reagents
- Autoclaved mortar and pestle
- Disposable plasticware suitable for RNA work (use in all post-lysis stages)
- RNA isolation kit (RNeasy, Qiagen)
- 2-mercaptoethanol
- 18–20 gauge needle and a disposable syringe
- Liquid nitrogen

Method

1. Collect worms by one of the methods discussed above (see *Protocol 1* or *2*) and concentrate into a slurry.

2. Estimate the volume of worms at this stage, and use this to calculate the appropriate volume of buffer needed below. Treat 0.25 ml of the concentrated worm slurry as for 250 mg of tissue if you follow the RNeasy protocol.

3. Freeze the collected worms rapidly and grind in liquid nitrogen

---

(exactly as in *Protocol 3*). Make absolutely sure, for RNA isolation, that the worms never warm up during the grinding process or prior to dissolution in denaturing buffer.

4. Disperse the frozen powdered worms in RNeasy RLT buffer to which 2-mercaptoethanol has previously been added, as per the manufacturer's protocol. Pass the lysate through an 18–20 gauge needle with a disposable syringe to aid dispersal and release of RNA.

5. Follow the remainder of the manufacturer's protocol.

---

[a] The method given here makes use of the commercial RNA isolation kit RNeasy manufactured by Qiagen. The details given here do not replace those provided with the handbook for the kit, but indicate relevant modifications. The relevant method is that described for total RNA isolation from animal tissues. To make this method as widely adaptable as possible, some suggestions of alternative methods that do not rely on this kit are made.

[b] *Alternative methods*: Other methods of RNA extraction can be applied. The guanidinium thiocyanate/caesium chloride cushion procedure described in ref. 1 works well, although it is more laborious than the commercial kits. As a general guide, the RNeasy RLT extraction buffer in step 4 above can be replaced with other denaturing RNA extraction buffers, suspending 1 volume of worms in 4–10 volumes of denaturing buffer, depending on the nature/concentration of the denaturing buffer used. A 1:4 dilution in TRIZOL from GIBCO BRL works for worms. If following the guanidinium thiocyanate/caesium chloride procedure (1), pass the lysate through an 18–20 gauge needle several times to shear genomic DNA, then centrifuge to remove any particulate matter prior to loading on to the caesium cushions.

[c] *Poly-A selection*: The RNA prepared from the above method, or its adaptations, is suitable for further mRNA purification using standard procedures. We routinely use the Oligotex kits by Qiagen.

For more rapid preparations, larvae and adults can be extracted directly without prior grinding in liquid nitrogen. A method that solubilizes intact worms in an SDS/Proteinase K-containing buffer is given in *Protocol 5* below, and an alternative TRIZOL-based method (GIBCO BRL) is also discussed. In my experience, the quick methods that do not involve grinding in liquid nitrogen do not work adequately for embryonic RNA. Although, for many purposes, these methods are adequate for extracting RNA from larvae and adults, the RNA may not be as intact or the yields not quite as great compared to that obtained by grinding in liquid nitrogen. The method in *Protocol 5* is ideal for generating RNA for semi-quantitative reverse transcriptase PCR (rtPCR) analysis as discussed in Section 4.4 below. The RNA is also of high enough quality for some Northern blot applications. This method is particularly useful where multiple samples have to be processed. It can be used in conjunction with the methods for generating synchronous cultures (see Chapter 4, *Protocol 7*), or with small populations of highly synchronous animals growing on plates. The method described in *Protocol 5* has been used for RNA time courses of post-embryonic development using synchronous cultures (see Chapter 4 for methods of generating synchronous cultures and Section 4.4.1 below). The method as described does not attempt to separate bacteria from worms prior to RNA preparation. This allows the method to be performed on very small numbers of animals without significant loss of material and it also permits rapid sampling, useful for time courses. If desired, the methods described in Section 2 of this chapter can be used to purify worms, in which case 1 volume of worm slurry should be added to 10 volumes of lysis buffer in *Protocol 5*, step 4. For all steps after the phenol/chloroform extraction (see *Protocol 5*, step 5), care should be taken to ensure that all reagents are RNase-free.

Protocol 5. Small-scale preparation of RNA by proteolytic digestion

Equipment and reagents

- Disposable plasticware, suitable for RNA work
- Lysis buffer: 0.5 ml 0.5% SDS, 5% 2-mercaptoethanol, 10 mM EDTA, 10 mM Tris–HCl pH 7.5, Proteinase K at a final concentration of 0.5 mg/ml (pre-heated to 55 °C)
- Phenol/chloroform (1:1) equilibrated to pH 5.2 with 10 mM sodium acetate, 50 mM NaCl, 3 mM EDTA
- Chloroform

- 3 M sodium acetate pH 5.2
- Ethanol
- RNase-free water
- 60 mM NaCl, 60 mM Tris–HCl pH 7.5, 60 mM MgCl$_2$, 10 mM DTT.
- RNase-free DNase I
- TRIZOL (GIBCO BRL)
- Disposable syringe fitted with an 18–20 gauge needle

A. *Standard method*

1. Wash worms off a single 5 cm plate with 1 ml of water and transfer to a microcentrifuge tube.

2. Centrifuge in a microcentrifuge for 30 sec to pellet the worms and bacteria.[a]

3. Pipette off the supernatant.

4. Resuspend the pellet in lysis buffer and incubate at 55 °C for 1 h.

5. Extract with 1 vol. of phenol/chloroform. Take care to avoid all the material at the interphase as much of the DNA will be present there.

6. Extract with 1 vol. of chloroform.

7. Precipitate the RNA by adding 0.1 volume 3 M sodium acetate pH 5.2 and 2.5 volumes of ethanol and cooling to –20 °C for 30 min.

8. Pellet the RNA in a microcentrifuge for 10 min.

9. Remove the supernatant and wash the pellet in 70% ethanol.

10. Resuspend the pellet in RNase-free water, typically 100 μl.

The following steps, to reduce the amount of contaminating DNA in the preparation, are optional, depending on the intended use of the RNA.

11. Adjust the RNA samples to 6 mM NaCl, 6 mM Tris–HCl pH 7.5, 6 mM MgCl$_2$, 1 mM DTT and treat with RNase-free DNase I (final concentration 50 μg/ml) at 37 °C for 1 h.

---

12. Re-extract with phenol/chloroform and chloroform, and ethanol-precipitate as in steps 7 to 10 above. Re-dissolve in 0.1 ml of RNase-free water.

B. *TRIZOL alternative method*[b]

1. Add 1 volume of worms to 9 volumes of TRIZOL and suspend rapidly either with vigorous vortexing or preferably by passing up and down through an 18–20 gauge needle at least 10 times, using a disposable syringe.

2. Incubate for 10 min with frequent agitation.

3. Pellet any remaining insoluble material by microcentrifuging for 10 min, and transfer the supernatant to a fresh tube.

4. Follow the remainder of the manufacturer's protocols without further modification for phase separation and precipitation of RNA.

[a] The presence of bacteria helps produce a robust pellet, so there is no need to cool the worms first.
[b] TRIZOL can also be used with intact worms.

The above methods give RNA suitable for most Northern blot analyses and rtPCR methods (see below). In both of the above methods, a certain amount of contaminating genomic DNA can be present. This is usually only a problem in Northern blots for particularly large transcripts, as the DNA tends to run as very high molecular weight material. It is much more of a problem for the semi-quantitative rtPCR method discussed below. If necessary, mRNA can be purified by standard poly-A selection procedures.

4.3 Northern blot analysis

Any standard method (1) of performing Northern blots will work with the *C. elegans* RNA prepared as above. The only worm-specific issue that should be considered is the multicellular nature of the animal and the inability of making tissue-specific RNA (see Section 4.1 of this chapter for a discussion). The abundance of a transcript that is present only in a few cells or for a short developmental period may be very low in a mixed-stage extract.

4.4 Semi-quantitative rtPCR

Semi-quantitative rtPCR is a very useful procedure for detecting changes in the steady-state abundance of mRNA species between different samples. It is particularly useful where many samples are to be processed, or where only small amounts of RNA can be obtained that are insufficient for Northern analysis. Examples of where this can be useful include comparisons between different mutant strains, differences in RNA prepared from specific developmental stages/developmental time courses (9), and between populations of animals that have been treated differently (e.g. exposed to different drugs, etc.).

4.4.1 Stage-specific RNA preparations

Semi-quantitative rtPCR can be a valuable tool for analysing differences in the temporal pattern of gene expression throughout the worm's life-cycle. Highly synchronous cultures of *C. elegans* can be generated for all stages of post-embryonic development (detailed in Chapter 4, *Protocol 7*). This method relies on a developmental arrest that occurs early in L1 larval development if purified embryos are hatched in the absence of food. However, good synchrony can only be maintained if the starved L1 larvae are cultured on standard *C. elegans* plates at relatively low density. It is essential that food never becomes limiting during the time course of the synchronous culture or otherwise synchrony is lost. Liquid cultures are inadequate as much greater synchrony is maintained on plate culture. This translates to about 10 000 larvae on a standard, 9 cm seeded plate. Considerably higher numbers can be used if only early-stage larvae time courses are required. A 9 cm plate can comfortably contain many more L1 larvae than adults. A single plate can be used for each time point required, generating sufficient RNA for most rtPCR type uses. There is little point in sampling more frequently than at 1-hour intervals because of the level of asynchrony within the synchronous culture technique. The RNA extraction methods given in *Protocol 5* are suitable for such time courses. This method provides synchronous animals covering larval and adult development suitable for RNA preparations for rtPCR time courses (9). At about 2 h after the L4 to adult moult, the first fertilizations occur giving rise to embryos *in utero*. Some methods of RNA preparation, such as those in *Protocol 5*, may not efficiently extract embryonic RNA once the eggshell is laid down. Nevertheless, the possible presence of embryonic RNA in preparations from adults must be considered. It may be possible to make use of this for time-course RNA from embryos. There is no easy way of obtaining sufficient synchronous embryos for RNA time courses. However, the adults from synchronous post-embryonic cultures do start to fertilize eggs relatively synchronously. They therefore contain populations of embryos that include progressively older stages, and this may be a way of addressing the timing of gene expression during embryogenesis. As detailed in Chapter 4 *Protocol 5*, mixed-stage embryos can be prepared by treating cultures with bleach. Once washed, RNA can be prepared from the embryos (see Section 4.2 of this chapter). All but the very earliest stages of embryogenesis can be represented in this way. Very early embryos are not resistant to the bleach treatment.

4.3.2 Contaminating genomic DNA

rtPCR relies on the use of reverse transcriptase to generate a cDNA copy of mRNA, followed by PCR using specific oligonucleotide primers to amplify specific cDNA sequences. Contaminating genomic DNA is also a target for such PCR reactions. For genes that contain introns, the presence of contaminating DNA does not always constitute a problem. Oligonucleotide primers for use in the PCR stage of the rtPCR reaction can either be designed to an intron/exon boundary such that genomic DNA is not a suitable template, or the primers can be designed to amplify across an intron such that cDNA-derived amplification gives a smaller product than amplification from genomic DNA. Many *C. elegans* introns are small, 50–100 bases being common. Frequently, primers may be designed that will permit the simultaneous amplification of both cDNA and genomic fragments such that the difference in size can be easily detected upon electrophoresis. This approach can be used to estimate the extent of DNA contamination, relative to the abundance of a particular RNA species. For some genes, the DNase treatment described in *Protocol 5*, steps 11 and 12, may be necessary. In the absence of DNase treatment, the genomic DNA-amplified fragment can be too abundant, relative to a low-abundance mRNA species, such that the mRNA species is not detected in the PCR reaction.

4.3.3 rtPCR quantification

Quantification with rtPCR can be achieved by a variety of methods. Depending on the purification procedure (see Section 2 of this chapter), *C. elegans* RNA is generally contaminated by varying amounts of *E. coli* RNA. Methods that attempt to quantify a given transcript in relation to the total RNA present should therefore be avoided. In the method described in *Protocol 6*, the simultaneous amplification of a control gene and the gene to be tested is performed. The relative abundance of the amplified, test-transcript fragment to that of the control gene generates a ratio that can be compared between different samples. Fluctuations in the relative abundance of the test transcript, as compared to that of the chosen control, are measured. The gene *ama-1* encodes the large subunit of RNA polymerase II and has been shown to be a valid control in this procedure, although it may not be a good control for early embryonic development.

  For the 'internally-controlled, semi-quantitative rtPCR' procedure described in *Protocol 6* cDNA is synthesized from the various RNA samples to be compared using Superscript-II obtained from Life Technologies. Random 9mers are used to prime the first-strand synthesis to ensure that the cDNA synthesized is not biased towards the 3' ends, biased against long messages, or biased against messages that have been partially degraded. The method of detection used relies on the PCR amplification of relatively short regions of cDNA, hence the starting RNA does not have to be completely intact. If using alternative RNA purification methods to that described in *Protocol 5* above, the Superscript-II methods should be consulted for the amounts of RNA to be used. Clearly, the method can be adapted for use with another source of reverse transcriptase. The method as described is designed to be used in conjunction with the small-scale SDS/protease digestion method set out in *Protocol 5*. Because of the contamination of *C. elegans* RNA with bacterial

RNA, it is not possible to be certain about the absolute amount of worm RNA present, but the volumes work with the yield obtained from that method as described.

The amount of cDNA mentioned in the protocols has been found empirically to work for the detection of the *ama-1* control transcript and other relatively abundant transcripts such as those of collagen genes (9). For low-abundance transcripts, it may be appropriate to use smaller dilutions of the starting RNA or to use undiluted cDNA in the PCR step.

The key issue to the PCR step is the appropriate design of oligonucleotide primers for amplification. The oligonucleotide sequences suggested below for the amplification of the control *ama-1* transcript generate a 522-bp fragment from the genomic copy of *ama-1* and a 426-bp fragment from the cDNA copy. These are readily distinguished by agarose gel electrophoresis. Another gene could be used as the control, but it is best to amplify a cDNA fragment of the test gene that is smaller than that amplified for the control. Primers designed to amplify the test and control genes must generate fragments of distinguishable sizes, should not form oligodimers with one another and must have a similar T_m. New primers should be tested for specificity and compatibility by performing a standard genomic DNA PCR reaction first. Obviously, sources of *Taq* polymerase and PCR reaction conditions other than those described could be used. The PCRs should have gone to completion with the reagents used up. For transcripts with levels changing considerably during development, this protocol is generally adequate. For a more accurate indication of changes in abundance, the number of cycles should be reduced to stop the reaction during the logarithmic phase of amplification.

Protocol 6.   Internally controlled semi-quantitative rtPCR

Equipment and reagents

- TE buffer: 10 mM Tris–HCl pH 8, 1 mM EDTA
- Random 9mers in TE buffer
- Superscript-II (Life Technologies)
- 5 × First-strand buffer (supplied with the Superscript II enzyme)
- RNase-free water
- 0.1 M DTT

- dNTPs (10 mM each of dATP, dCTP, dGTP, dTTP)
- Control primers; *ama-1*-sense (CAGTGGCT-CATGTCGAGT) and *ama-1*-antisense (CGA-CCTTCTTTCCATCAT)
- Gene-specific primers
- *Taq* polymerase and buffer (Promega)

A. *cDNA synthesis*

1. Mix, in a microcentrifuge tube, 0.1 µg random 9mers dissolved in TE buffer, 2 µl of RNA (from the RNA generated in *Protocol 5* above), and sterile distilled water to a total volume of 11 µl.

2. Heat to 70°C for 10 min and then quickly chill on ice.

3. Centrifuge the contents briefly in a microcentrifuge for 5 sec.

4. Add 4 µl of the 5 × First-strand buffer, 2 µl 0.1 M DTT, 1 µl mixed

---

dNTPs (10 mM each), 1 μl of RNase-free water, and 200 units of Superscript-II to each reaction and incubate at 37 °C for 15 min followed by 42 °C for 1 h.

5. Stop the reaction by heating to 65 °C and dilute the sample by adding 400 μl of TE buffer to each reaction.

B. *rtPCR reactions*

1. Assemble, on ice, a 25 μl reaction containing 1 μl cDNA, 75 ng of each of the four primers, 200 μM dNTP, 100 ng/μl BSA, *Taq* DNA polymerase buffer containing 2.5 mM $MgCl_2$ and 2.5 units *Taq* polymerase.

2. Incubate the reaction at: 94 °C for 3 min (initial denaturation), followed by 36 cycles of 60 °C for 30 sec, 72 °C for 30 sec, and 94 °C for 30 sec, and finishing with 60 °C for 3 min, 72 °C for 3 min.

3. Examine the products by standard agarose gel electrophoresis (1)

The products of the rtPCR reactions are visualized by agarose gel electrophoresis using an appropriate concentration of agarose for the size of fragments to be resolved. In the thermocycling conditions described above, reactants such as dNTP should be depleted by the end of the reaction. If the test transcript is of a comparable abundance to that of the control and the primers chosen for both the control and test transcripts work equally well in the PCR reaction, the relative intensity of ethidium-stained bands on electrophoresis should be roughly equivalent. If the abundance of one mRNA species is considerably different from the other in the starting material, a difference will be seen in the relative intensities of amplified bands at the end of the reaction due to depletion of available dNTPs by the more abundant product. Furthermore, if the same primers and PCR conditions are then used for different samples of cDNA, differences in the relative abundance of test transcript to control can be seen. This does not provide an absolute measure of any transcript level in a sample, but it does detect changes between different samples as differences in the relative intensity of ethidium-stained bands. This method has been very useful when combined with RNA samples prepared from distinct life-cycle stages (9).

A more accurate quantification of changes in relative abundance is possible if the rtPCR reactions are stopped during the logarithmic phase, before intense bands can be detected with ethidium staining. Generally 25–27 PCR cycles are sufficient, but this must be checked empirically. Samples are electrophoresed in agarose gels, Southern blotted to membrane, and detected by probing (e.g. with radiolabelled probes, such as end-labelled primers) (9). Any suitable method of measuring the signal from the test- and control-amplified bands can be employed (e.g. scintillation counting of bands cut out from the Southern blot membrane or the use of a phosphoimager). The ratio of the test to control value can then be compared for different samples. It is possible to avoid blotting by employing ABI sequencing reagents in the PCR amplification stage and running the reactions on an ABI sequencer. The ABI software allows the relative quantification of the test and control bands detected in this way (Mark Blaxter, personal communication).

Microarray technology is being developed for *C. elegans* and will have a major impact on transcript analysis. It is anticipated that arrays will be available soon with all the predicted genes in the *C. elegans* genome represented. This technology is changing rapidly, at the time of writing, and so will not be described here. A source of current information should be found at (`http://cmgm.stanford.edu/_kimlab/`).

4.5 Transcript analysis

The structure of gene transcripts from *C. elegans* can be investigated in exactly the same ways as for other eukaryotes (1), but there are some points specific to *C. elegans*. Standard techniques such as Northern blot analysis to confirm the predicted size of a transcript, determination of cDNA sequence to confirm intron/exon structure, and the mapping of 3' and 5' ends should be used where appropriate. However, the intergenic sequence in *C. elegans* is A/T-rich with runs of A being relatively common, and oligo-dT priming from such sequences can sometimes occur. Therefore, when cDNA sequencing or 3' RACE is used to determine the 3' end of a transcript, the inferred site of poly-adenylation should be checked with the genomic sequence.

The most unusual aspect of gene structure in *C. elegans* is the existence of operons and *trans*-splicing (see Section 4.1.3 above). 5' RACE can be used to determine whether a gene is *trans*-spliced. The sequence of the *trans*-spliced leaders (see Section 4.1.3) can be used in oligonucleotides that are to function as the 5' primers in RACE reactions (in place of the adapter primer). cDNA can be generated from RNA using either gene-specific primers, or random primers, and PCR can be performed with nested, antisense, gene-specific primers and one or other SL primer. Any standard cDNA synthesis method can be used, either priming cDNA synthesis with random primers, or with a gene-specific primer (see *Protocol 7*). Used in this way, the 5' RACE protocol will indicate whether transcripts for a given gene can be detected *trans*-spliced to either of the SL1 or SL2 classes of *trans*-spliced leaders. It is very important to remember that this is *all* that it indicates. This approach will obviously not detect non-*trans*-spliced transcripts, and, because of the extreme sensitivity of the method, it cannot be assumed that any detected *trans*-spliced transcript necessarily represents the most abundant mature transcript of the gene. For this reason, the use of a standard 5' RACE protocol, such as those available in kit form (e.g. from GIBCO BRL), will provide a more complete analysis of the 5' end of a gene. The standard method does not rely on the presence of a *trans*-spliced leader for the PCR stage of the reaction, and can detect all transcript ends. Instead, it relies on the homo-polymer tailing of the 3' end of the first-strand cDNA (5' end of the gene) using terminal transferase. An adapter primer that hybridizes with the synthetic homo-polymer tail added with terminal transferase is used as the upstream primer in second-strand cDNA synthesis. The PCR reactions should use a smaller variant of the adapter primer that does not contain the homo-polymer region. Both of these are provided with the GIBCO BRL 5' RACE kit. Similar considerations regarding gene-specific primer design for use in the PCR stages (or for gene-specific cDNA synthesis) apply to the general and SL methods. In both cases, amplified bands can be cut from a gel and sequenced directly using gene-specific primers (we regularly use the same primers used in the PCR reaction, or nested primers can be used). The amplified bands can also be cloned then sequenced. If the standard method is used, the sequence of the amplified fragment will indicate the presence or absence of an SL sequence at the 5' end of the transcript.

   The following should be considered when designing gene-specific primers for RACE procedures. At least two nested, antisense, gene-specific primers will be needed. Either the downstream primer will be used for cDNA synthesis and a single PCR reaction performed with the nested and SL primers (or adapter primers), or random primers will be used for cDNA synthesis and two rounds of PCR performed. The latter approach can be useful if several genes are to be analysed. (In rare cases, a third gene-specific primer may be necessary). The antisense, gene-specific primers should be designed taking into consideration the length of the predicted products (shorter lengths may be more efficiently generated both in the cDNA synthesis and PCR steps). Primer specificity can be considered by comparison with the entire *C. elegans* genomic sequence (BLAST searches, etc.), and primers to be used in the PCR steps should be designed to avoid the formation of oligodimers with all primers to be used in the reactions.

Protocol 7.   5' transcript analysis with RACE using the SL1 and SL2 sequences

Equipment and reagents

- Superscript-II (Life Technologies, although other sources of reverse transcriptase could be used)
- First-strand buffer (supplied with the enzyme)
- Primers in TE buffer (see *Protocol 6*)

- Sterile, distilled RNase-free water
- 0.1 M DTT
- dNTPs (10 mM each of dATP, dCTP, dGTP, dTTP)
- Agarose gel electrophoresis equipment and reagents

Method

1. Mix, in a microcentrifuge tube, 0.1 μg of random 9mers or downstream gene-specific, antisense primer dissolved in TE buffer, 0.5 μg of poly-A selected RNA, and sterile, distilled RNase-free water to a total volume of 12 μl.

2. Heat to 70°C for 10 min and quickly chill on ice.

Protocol 7. *Continued*

3. Briefly centrifuge the contents to the base of the tube (5 sec in a microcentrifuge).

4. Add 4 μl of the 5 × First-strand buffer, 2 μl 0.1 M DTT, 1 μl dNTPs, and 200 units of Superscript-II.

5. Incubate at 37 °C for 15 min followed by 42 °C for 1 h.

6. Stop the reaction by heating to 65 °C for 15 min.

7. Perform two separate PCR reactions (one for each of the two SL oligos) plus a nested gene-specific primer.[a]

8. Electrophorese a fraction of the PCR product on an agarose gel and visualize by ethidium staining.

9. Perform a second, nested PCR reaction with another gene-specific primer on a dilution from the first PCR reaction if no well-defined bands are present.

10. Purify bands cut from the gel and sequence directly by standard thermocycling methods—use either the primers used for the amplification or ones designed internal to the fragment. Alternatively, clone and sequence the products.

[a] 1 μl of cDNA from the above reaction should be sufficient. In some cases it may be possible to use greater dilutions. Optimal conditions for PCR will vary depending on the design of the gene-specific primers, the length of the amplified product, abundance of the transcript, etc. Test this empirically.

For those genes that are *trans*-spliced, it may be desirable to determine the start of the primary transcript (i.e. the pre-mRNA containing the RNA sequence that is removed from the 5′ end of the transcript during the addition of the SL *trans*-spliced leader). Obviously, a standard 5′ RACE protocol such as the GIBCO BRL 5′ RACE kit must be used. Total RNA (5 μg) should be used for cDNA synthesis as pre-mRNA should be absent from the poly-A fraction. cDNA synthesis should be performed as indicated in the kit protocol, and tailed. It is advisable to use a gene-specific primer to prime cDNA synthesis, designed to hybridize relatively close to the 5′ end of the mature SL-containing transcript. To detect the pre-mRNA sequence upstream of the SL addition site, an antisense, gene-specific primer must be designed upstream of the site of SL addition for use in the nested PCR step (in some cases, two nested upstream primers may be needed). A first round of PCR is performed with the general adapter primer(s) and an antisense, gene-specific primer that will amplify both the mature SL end and the upstream pre-mRNA end. Invariably, the pre-mRNA end, presumably representing the true start of transcription, is in much lower abundance than the *trans*-spliced end. A nested PCR reaction is performed on a fraction of the first PCR reaction, using the gene-specific primer from upstream of the SL site. This method can only work if the transcriptional start site is far enough upstream of the SL addition site to permit the chosen upstream gene-specific primer to prime. This method has been used to map the pre-mRNA end for a gene with a relatively abundant transcript (10). Due to the necessity of using total RNA (pre-mRNA will be absent from poly-A selected RNA), and the very short life of pre-mRNA, the target molecules will be present in very low abundance. The use of nuclear RNA instead of total RNA may increase the sensitivity of this method, if necessary.

5. Protein detection

Methods of protein extraction are generally more dependent on the biochemistry of the particular protein of interest than on anything that relates to the specific organism from which they are extracted. This is a field in itself and cannot be dealt with in depth here. Some general considerations that relate specifically to *C. elegans* and simple methods of general protein extraction will be discussed. The *in-situ* detection of protein by immunofluorescence techniques in *C. elegans* is dealt with in Chapter 9.

As with RNA extraction, the life-cycle stage of expression is important when considering proteins. It is useful to know when (and possibly where) the transcript of a gene is expressed before considering the extraction of its encoded protein.

5.1 Simple, whole animal lysates

The simplest method of extracting protein from *C. elegans*, suitable for purposes such as Western blot analysis, is simply to harvest worms as for RNA or DNA, taking steps to reduce bacterial contamination, and to pellet the worms by centrifugation. The volume of the concentrated slurry of worms is estimated, diluted directly in standard reducing SDS–PAGE sample buffer (11) including marker dyes (1 volume of worms to at least 5 volumes of sample buffer), and boiled for 5 min. Samples are centrifuged briefly in a microcentrifuge to pellet non-solubilized worms and then the supernatant is loaded directly on to a reducing SDS–PAGE gel. Other texts should be consulted for details and conditions for PAGE (12). This procedure should be performed immediately before electrophoresis and not used where protein has to be stored. This method is not suitable for non-reducing conditions. In the absence of a reducing agent, worms do not dissolve and hence the release of cellular proteins may be limited. It is unsuitable for extractions from embryos.

5.2 Protein extraction methods

As with nucleic acid extraction, opening the worms is a key issue. For many purposes, separation of cellular proteins from the extracellular cuticle is desirable. The nematode cuticle contains abundant collagenous proteins that constitute a significant percentage of the total protein. They are largely insoluble in the absence of reducing agents, permitting various detergent extractions to be performed without solubilization of these collagens. This insoluble cuticular material can be removed by centrifugation. A method of separating cellular from cuticular material is given in *Protocol 8*.

For small-scale extractions, small homogenizers designed to fit a microcentrifuge tube can be quite effective. Larger scale extractions can be achieved using tight-fitting glass homogenizers or by sonication. In each case the extraction buffer should be kept chilled with ice. The resilient nature of the worm cuticle requires extensive homogenization/sonication to obtain good yields of internal proteins. The presence of detergents in the extraction buffer certainly helps. Even after considerable periods of homogenization the cuticles usually appear intact. I suspect most of the cellular material that is extracted by these procedures exits the cuticle through openings at the mouth, anus, or vulva. There is a very obvious change in the optical appearance of worms when the cellular contents have been extracted. When trying an extraction method for the first time, it is advisable to observe small numbers of animals under a microscope after various periods of homogenization or sonication. The 'ghosts' of extracted cuticles are very obvious when compared with animals that have not been successfully extracted.

The liquid-nitrogen grinding method of fragmenting worms (see *Protocol 3*) is very efficient at quickly breaking open worms and it is also successful for embryos. This method can be adapted for protein extractions; however, any method that includes freezing will cause the liberation of lysosomal proteases. It is therefore necessary to include protease inhibitors. The choice will be dependent on the nature of the protein to be extracted.

Protocol 8. Separation of cellular proteins from cuticular proteins[a]

Equipment and reagents
- TBS: 0.05 M Tris pH 7.6, 0.15 M NaCl
- *n*-octyl glucoside
- Protease inhibitors[b]

Method

1. Collect worms by filtration or centrifugation (see *Protocol 1*), washing once in TBS and collecting by centrifugation at 10 000 *g* for 5 min.

2. Resuspend in 5 volumes of TBS, 1.5% *n*-octyl glucoside, plus protease inhibitors[b, c].

3. Homogenize for 10 min, then check for the release of cellular material by the appearance of 'ghost-like' cuticles.

4. Incubate on ice for 30 min.

---

5. Centrifuge at 10000 *g* for 30 min at 4°C and collect the supernatant which contains cellular proteins.

6. Store at –70°C or use as appropriate.

7. If cuticular material is desired, extract the pellet from step 4 by solubilizing in reducing SDS–PAGE sample buffer with boiling for 5 min. Centrifuge at 10000 *g* for 15 min and remove the supernatant for PAGE.

[a] Consideration should be made of the removal of bacteria from the worms and the issue of the life-cycle stages present in the culture used (see Section 2 of this chapter). This method is a personal communication from Tony Page.
[b] The choice of protease inhibitors is dependent on the sensitivities of different proteins, but could be: PMSF (1 mM), EDTA (1 mM), EGTA (1 mM), E-64 (2 µM), or Pepstatin-A (0.1 µM). (All available from Sigma.)
[c] For some proteins, protease inhibitors are not essential if everything is done fast and kept cold.

References

1. Sambrook, J., Fritsch, E. F., and Maniatis, T. (1989). *Molecular cloning: a laboratory manual* (2nd edn). Cold Spring Harbor Laboratory Press, NY.
2. McPherson, M. J., Quirke, P., and Taylor, G. R. (ed.) (1991). *PCR: a practical approach*. IRL Press, Oxford.
3. Wood, W. B. (ed.) (1988). *The nematode* Caenorhabditis elegans. Cold Spring Harbor Laboratory Press, NY.
4. Riddle, D. L., Blumenthal, T., Meyer, B. J., and Priess, J. R. (ed.) (1997). C. elegans *II*. Cold Spring Harbor Laboratory Press, NY.
5. Cline, T. W. and Meyer, B. J. (1996). *Annu. Rev. Genet.*, **30**, 637.
6. Hodgkin, J., Horvitz, H. R., and Brenner, S. (1979). *Genetics*, **91**, 67.
7. Klass, M. and Hirsh, D. (1981). *Dev. Biol.*, **84**, 299.
8. Blumenthal, T. (1995). *Trends Genet.*, **11**, 132.
9. Johnstone, I. L. and Barry, J. D. (1996). *EMBO J.*, **15**, 3633.
10. Gilleard, J. S., Barry, J. D., and Johnstone, I. L. (1997). *Mol. Cell. Biol.*, **17**, 2301.
11. Laemmli, U. K. (1970). *Nature*, **227**, 680.
12. Hames, B. D. (ed.) (1990). Gel electrophoresis of proteins: a practical approach (2nd edn). IRL Press, Oxford.

11

Biochemistry of *C. elegans*

PAUL E. MAINS and JAMES D. McGHEE

1. Introduction

Notwithstanding the virtues of *C. elegans* as an experimental system and extolled in the earlier chapters of this book, it is probably only fair to say that *C. elegans* is not an ideal organism in which to study biochemistry. It is tedious to grow and disrupt worms in the large quantities that are the usual starting point for conventional biochemistry, at least for the isolation of rare molecules. It is also difficult to produce tightly synchronized populations on a large scale—eggshells and cuticles present impressive barriers—and it is essentially impossible to produce biochemically useful quantities of pure tissues. There are two exceptions to this last statement, i.e. the production of sperm (1) and the production of oocytes (2, 3; and see below).

With the genome project-stimulated influx of workers from other fields into the study of *C. elegans*, we can expect to see an increased emphasis on biochemical approaches. Many of these workers have used biochemistry to study their favourite gene product in higher organisms and are likely to continue this approach in their study of worms. Since the justification of the present volume is to collect techniques that would be useful to such newcomers, we describe several common procedures aimed at more biochemical-type experiments. We concentrate on techniques we have used in our own laboratories, where there is an emphasis on early embryonic development. As described in Chapter 3, the World Wide Web has made a great variety of technical descriptions easily accessible.

2. Large-scale growth of mixed-stage *C. elegans* on solid media

Nematodes used in pest control are routinely grown in boxcar-sized fermentors, but *C. elegans* is rarely (if ever) grown on this scale. The majority of molecular biology techniques are satisfied with a gram or two of worms. For more biochemically oriented projects, we have grown *C. elegans* on a scale of several hundred grams at a time, adequate for the purification of proteins present at reasonable abundance in the worm population. *C. elegans* can be grown at this scale in either liquid culture or on solid media. Here, we describe only growth on solid media, as we feel this is a more convenient and robust procedure for the starting investigator. Growth on solid media is flexible (the use of mutant strains presents little problem), and does not require specialized equipment (beyond standard temperature-regulated incubators). Although there is no doubt that liquid-grown worms are cleaner, they also appear starved. In our own experience, we have found growth in liquid to be somewhat unreliable. Others do not share this view and methods to grow *C. elegans* in liquid culture have been described in Chapter 4 (or see ref. 4).

The basic protocol we use to produce 10–200 grams of *C. elegans* is growth on 'egg plates'. This has evolved from a recipe developed years ago by Raja Rosenbluth and David Baillie (Simon Fraser University). We have used mixed populations of wild-type worms produced by this protocol to purify the digestive enzymes GES-1 (5) and PHO-1 (6) to homogeneity (5000–10000-fold purification) as well as to produce microtubules for exploring interactions with proteins of the meiotic apparatus. The basic protocol can be used as the starting point to produce large quantities of dauer larvae, which can then be used, after feeding and further incubation, to produce staged L4 and adult worms. From gravid adults, gram quantities of mixed-stage embryos can be obtained. By hatching in the absence of food, these embryos produce a synchronized population of L1 larvae, which, upon feeding and further incubation, produce synchronized populations of later larval stages. By using worms containing a temperature-sensitive mutation in the *fer-1* gene (7), we have produced several grams of unfertilized oocytes (2, 3). By using other mutant strains, one can obtain adult populations lacking sperm and fertilized embryos (*fem-1*), lacking oocytes and embryos (*fem-3(gf)*), or lacking total germline (*glp-4*); with various combinations of these strains, it can be determined whether an mRNA or protein is expressed in the soma, mitotic germline, sperm, oocytes, or fertilized embryos. Finally, by incorporating fluorodeoxyuridine into the medium, we have produced large quantities (tens of grams) of a synchronized population of embryos blocked in mid-proliferation stage. These blocked embryos have been used to produce embryonic nuclei and embryonic nuclear extracts to investigate the presence of gene-specific binding proteins (2, 8).

The following protocol is calculated for four trays at a time, with 350 ml of medium per tray.

Protocol 1. Large-scale growth of *C. elegans* on egg plates

Equipment and reagents
- Instrument/pipette sterilizing pan (Nalgene, cat. no. 6910–0618)[a]
- 1 M KPO$_4$ pH 6
- 1 M CaCl$_2$
- 1 M MgSO$_4$
- 5 mg/ml cholesterol in ethanol

- Standard incubators for *C. elegans* growth[b]
- Standard Waring blender with 500 ml autoclavable glass containers (screw lids are an advantage)
- NGM (Nematode growth medium) and standard buffers (e.g. M9 and PBS, see Chapters 4 and 9, respectively)[c]
- Magnetic stirrer and bar
- Extra-large chicken eggs
- 70% ethanol
- Sterile pipettes
- Standard 2 × YT broth

- Wild-type *E. coli* culture (e.g. N99; originally from M. Gellert, NIH)
- Bent glass rod
- 50 ml polypropylene, screw-cap, centrifuge tubes
- 50% sucrose (w/v) in water
- Swinging bucket rotor
- Wide-bore Pasteur pipette (or P5000 Pipetman)
- Dewar flask (or 50 ml plastic centrifuge tube)
- Liquid nitrogen and safety equipment

NB: All equipment (trays, beakers, stir bars, cylinders, blender jars, centrifuge bottles, etc.) should be autoclaved ahead of time.

A. *Preparation of NGM–0.5 × egg trays*

1. Mix 33.75 g of NGM agar (4.5 g NaCl, 25.5 g agar, and 3.75 g peptone) with 1.5 litres of distilled water. Autoclave for 20 min at 20 lb/inch^2 and let the agar cool to 50–60 °C.

2. Add 37.5 ml 1 M KPO$_4$, 1.5 ml 1 M CaCl$_2$, 1.5 ml 1 M MgSO$_4$, and 0.75 ml of the cholesterol solution. Pour about 350 ml/tray into four trays, cover and leave to harden completely. Increase the peptone concentration two- to threefold to achieve a thicker bacterial growth.[d]

3. For sufficient egg mixture for four trays (50 ml/tray), place 100 ml of sterile water (along with a sterile, magnetic stirring bar) into a 500 ml (autoclaved) beaker. Microwave until just boiling.

4. Sterilize the surface of two extra-large chicken eggs with 70% ethanol and break the eggs into the water, rupturing the yolks with a sterile pipette.

5. Stir on a hot plate at medium heat for 5 min.

6. Transfer the heated egg mixture (but not the stir bar) to a glass blender jar (with screw lid) and mix for 2 min at the highest setting.

7. Leave it to settle for about 15 min and then spread approximately 50 ml evenly over the surface of each tray. Flame the surface briefly to remove bubbles and leave to dry without the lids in a flow hood until the surface is tacky, more-or-less the consistency of tofu. Do not overdry.

B. *E. coli and worm inoculation*

1. Grow an overnight *E. coli* culture[e] at 37 °C in standard 2 × YT broth.

2. Distribute 300 ml aliquots of this culture into 500 ml centrifuge bottles and spin at 4500 *g* for 5 min at 4 °C.

Protocol 1. *Continued*

3. Resuspend each pellet in about 10 ml of 2 × YT broth and spread on each tray using a bent glass rod. (Alternatively, the bacteria can be mixed with the above egg mixture, once cooled and prior to spreading.) Incubate the trays at 25 °C overnight to make the bacterial growth more uniform, but this is not necessary.

4. Inoculate each tray with a small number of worms washed from a standard 60 mm NGM stock plate. Allow the population to grow through multiple cycles and monitor growth daily by scooping up a small sample with a sterile spatula or wormpick for microscopic observation.[f]

C. *Harvesting the worms*

1. Harvest the worms when the population is reaching saturation (or is not changing with time). Add 100 ml distilled water to each tray and gently agitate at room temperature for 5–10 min.

2. Transfer the worms and debris to 500 ml centrifuge bottles. Rinse the trays with water and combine with the original harvest. Use cold water or buffer to minimize the movement of the worms; this results in a tighter pellet.

3. Pellet the worms by centrifuging at 1600 g for 5 min at room temperature.

4. Remove this (and all) supernatants by aspiration.

5. Wash worms a further 2–3 times with water to remove some of the smaller debris, i.e. most of the bacteria will not pellet.

D. *Cleaning the worms by sucrose flotation*

1. Resuspend the pooled worms (still from four trays) in 120 ml of cold water and divide into six, 50 ml polypropylene, screw-cap, centrifuge tubes, i.e. 20 ml of worms each.

2. Add sterile 50% sucrose (w/v in water) to a final volume of 50 ml per tube **and mix gently**.

3. Spin in a swinging bucket rotor for 5 min at 50 g in the cold, followed by a second spin at 1000 g for a further 5 min. (This double-spin protocol separates the worms (which float) from bacterial and agar debris (which sinks).)

4. As soon as possible after centrifugation, use a wide-bore Pasteur pipette or a P5000 Pipetman to collect worms from the top of the tubes and transfer them to 500 ml centrifuge bottles. Wash three times as above with water.

5. After the final wash, resuspend the pelleted worms in an equal volume of PBS or M9 buffer.

E. *Long-term storage of worms in liquid nitrogen—optional*

1. Incubate the resuspended worms at room temperature for 30–60 min with gentle agitation in order to allow digestion of intestinal contents.

2. Wash the worms one final time and resuspend in an equal volume of PBS or M9 buffer.

3. Use a wide-mouth Pasteur pipette to dribble the worm suspension into liquid nitrogen in a Dewar flask or, for small volumes, into a 50 ml plastic centrifuge tube. (The drops immediately freeze into small pellets.) Add the worms slowly so that the drops freeze without sticking to one another; add more liquid nitrogen as needed.

4. Allow the liquid nitrogen to evaporate before sealing the tube (to avoid explosion).

5. Store at −70 °C.[g]

---

[a] These polypropylene trays are autoclavable, stackable, and have a fitted lid. Each tray (456 × 152 × 67 mm) uses about 350 ml of medium and can yield 5–10 grams of worms. We grow between 1 and 24 of these trays at a time. If such trays are unavailable, standard Pyrex baking dishes (covered with aluminium foil for autoclaving and plastic wrap for growth) can be used (4).

[b] We have grown various mutant strains at temperatures ranging from 15 °C to 26 °C. For a temperature-sensitive strain such as *fer-1(b232)*, it is important to be able to hold the temperature at 26 °C with a precision of a few tenths of a degree.

[c] For washing and storing worms and embryos, we tend to use M9 and PBS interchangeably.

[d] It is a common concern that plates made with bacterial-plate grade agar contain substances that inhibit restriction enzymes, etc. If the worm stages are to be used for purposes involving nucleic acid modifying enzymes, we use NGM trays made with agarose instead of agar or (less expensively) agar trays overlaid with several millimetres of agarose-containing medium.

[e] Wild-type *E. coli* produces a more luxuriant food source than does the commonly used OP50 strain, a uracil auxotroph. We use a strain designated 'N99', originally obtained from M. Gellert (NIH).

[f] In practice, when we grow worms on a regular basis for stockpiling, we seed trays from a population of dauer larvae, harvested and stored as described above. We inoculate each tray with about 1.5 million dauers and then incubate at 20 °C for several days. The health and stage of the population is monitored daily.

[g] Methods for disrupting the worms will be discussed below.

Depending on the time of harvest, the above protocol will yield a mixed-staged population. However, L1 larvae tend to be lost in the sucrose flotation step and will consequently be under-represented.

3. Production of dauer larvae

For many purposes, it is convenient first to let a batch of worms proceed into the dauer stage; dauers are then purified from any remaining non-dauer stages by treatment with SDS (9). Dauers provide a convenient and reproducible means to seed further growths, to produce synchronized L4 and adult stages, and to produce oocytes using *fer-1(b232)* worms, in which the temperature-sensitive period is after the dauer stage. *Protocol 2* describes a method for the initial production of a stock of dauer larvae.

Protocol 2.   Initial production of a stock of dauer larvae[a]

Equipment and reagents

- Starved, asynchronous, 60 mm stock plate of worms
- PBS or M9 buffer (see *Protocol 1*)
- NGM–0.5 × egg trays (see *Protocol 1*)
- Equipment and reagents for harvesting and cleaning worms (see *Protocol 1*)
- 1% SDS

Method

1. Wash worms of the desired genotype off a starved, asynchronous, 60 mm stock plate using water, PBS, or M9 buffer.

2. Resuspend in a small volume (e.g. 0.5 ml) and add them to NGM–0.5 × egg trays as described in *Protocol 1*, Part B.

3. Incubate the trays for 7–10 days at 20 °C.

4. Check for dauers by observing a sample under a dissecting or compound microscope. (Dauers are dark and very thin compared to non-dauer stages.) Incubate the trays until more than 90% of the worms have become dauers.

5. Harvest and wash the dauers as described in *Protocol 1*, Part C, including the dauers that sometimes congregate on the sides of the tray and which appear as a light-brown crust.

6. Kill non-dauers by resuspending the pellet (from four trays) in 100 ml of 1% SDS (the non-feeding dauers survive because they do not ingest the SDS). Gently rock the tube at room temperature for 15 min and then wash 2–3 times in cold water.

7. Proceed to the sucrose flotation step (see *Protocol 1*, Part D).

[a] The final yield is about 4 million dauers per tray.

---

4. Production of *C. elegans* adults and embryos

Embryos are isolated from gravid hermaphrodites using an alkaline hypochlorite solution (9), but such embryos are not synchronized. Hermaphrodites, especially when they are older or when the plates become starved, can retain their eggs so that embryos of almost all ages are likely to be present. Only those embryos younger than 30 min post-fertilization, for which the eggshell is not yet resistant to the treatment, will be absent from these preparations. With our usual protocols, most embryos will be less than 6-h post-fertilization, i.e. they will be in proliferation–pre-morphogenesis stages. If greater embryonic synchrony is desired, one can carefully watch the L4 population as it emerges from the dauer stage (see below) and harvest the worms when fertilized embryos first appear in the culture (11). While this variation sacrifices yield, a population of embryos can be obtained, most of which are pre-gastrulation (i.e. less than 28 cells and less than 2-h post-fertilization). Incubation of these early embryos in a sterile buffer solution for the appropriate time results in a population of embryos synchronized to 1–2 h around the desired stage.

Protocol 3. Production of *C. elegans* embryos

Equipment and reagents

- 1.5 × 10^6 dauers (see *Protocol 2*)
- NGM–0.5 × egg tray (see *Protocol 1*, Part A)
- Equipment and reagents for harvesting and cleaning worms (see *Protocol 1*, Parts C and D)
- 50 ml, screw-cap, polypropylene centrifuge tubes
- Equipment for Nomarski microscopy (see Chapter 7)

- Alkaline hypochlorite solution: 10 ml reagent-grade sodium hypochlorite solution (6% available chlorine), 2.5 ml 10 M NaOH, 22.5 ml water. Use only stock hypochlorite bottles that have been opened for less than 1 month and that have been stored at 4°C.
- M9 buffer or PBS (see *Protocol 1*)
- Seeded NGM plate

Method

1. Add 1.5 million dauers to each NGM-0.5 × egg tray and incubate at 20°C for 2–3 days. Check a sample under the microscope to ensure that all animals have exited the dauer stage and are now full of embryos.

2. Harvest and wash (including the sucrose flotation step) as described in *Protocol 1*, Parts C and D. Store gravid adults at this point, if desired.

3. Combine 1 volume of pelleted worms with 2.5 volumes of alkaline hypochlorite solution in a standard 50 ml, screw-cap, polypropylene tube. Make sure that the total volume of worms plus alkaline hypochlorite per 50 ml tube is < 25 ml.

4. Incubate for 4 min at room temperature with gentle mixing.

5. Fill the tube with M9 buffer (or PBS) and quickly pellet by centrifuging at 1000 *g* for 1 min. Replace the supernatant with the same volume of alkaline hypochlorite solution as before.

6. Carefully monitor the worms under the dissecting microscope, either directly in the 50 ml tube or by taking a small sample to a depression slide. When approximately half of the hermaphrodites have broken (they break in half at the vulva, after roughly 10 min in alkaline hypochlorite), fill the tube with M9 buffer and quickly centrifuge and wash three more times.

7. Float the embryos on sucrose and wash (see *Protocol 1*, Part D).

Protocol 3. *Continued*

8. Assess the developmental stage of the population by Nomarski microscopy.
9. Check viability by incubating a sample overnight on a seeded NGM plate; viability should be > 70%.

One tray of gravid hermaphrodites will yield more than 10^7 embryos. For long-term storage, packed embryos are diluted with an equal volume of PBS or M9 buffer and dripped into liquid nitrogen as described in *Protocol 1*, Part E.

5. Production of different stages of the *C. elegans* life-cycle

Reasonable quantities (grams) of staged worms can be produced by variations of the above protocols. This is usually more than adequate for producing, for example, staged RNAs for Northern blottings. Starting with embryos isolated as in *Protocol 3*, synchronized L1 larvae can be isolated by gently shaking the embryos overnight at room temperature in M9 buffer; to ensure adequate aeration, keep the surface area as large as possible. Hatched larvae remain healthy but do not grow in the absence of food. L1 larvae do not float on the usual sucrose solution (and are generally clean in any case) so are either frozen directly for long-term storage or are allowed to develop into L2 and L3 stage larvae—add 10^7 L1 larvae per seeded NGM-0.5 × egg tray and incubate for 24 h at 15 °C to produce L2 animals or for 24 h at 20 °C to produce L3 animals (10). Populations later than L3 are more tightly synchronized if produced from dauers rather than from L1 larvae. To each NGM-0.5 × egg tray, add 1–2 million dauers and incubate at 20 °C. Collect L4-stage larvae 1 day later and gravid hermaphrodites 2–3 days later. Animals are harvested by sucrose flotation as described in *Protocol 1*, Part D. All stages should be checked microscopically. One tray should produce about 10 ml of packed gravid hermaphrodites.

6. Production of embryos blocked in mid-proliferation phase by growth on fluorodeoxyuridine (FUdR)

One of the central projects in our laboratories has been to study the transcriptional regulation of a gut-specific esterase gene named *ges-1* (2, 8, 12–14). Our approach has been more biochemical than genetic, starting with the analysis of the *ges-1* promoter and then proceeding to the identification of upstream activating genes (8, 14). The simple, but powerful, techniques of electrophoretic mobility shifts and footprinting have rarely been applied to *C. elegans* extracts. This is unfortunate, since there are likely to be many important regulatory genes whose products can be detected biochemically but would be difficult to identify by the usual methods of classical genetics (for example, if the mutations are lethal).

We have previously described an experimental system that has allowed us to identify sequence-specific binding factors directly in embryonic nuclear extracts (2, 8). We have found that growing parent *C. elegans* worms in the presence of both the deoxynucleoside analogue fluorodeoxyuridine (FUdR) and thymidine leads to complete blockage of the next generation embryos in the mid-proliferation stage of development, presumably because of limitations in the supply of DNA precursors. The blocked embryos are quite homogeneous (200 ± 40 C values of DNA), remain viable for days, and can be easily produced in amounts (tens of grams) that are adequate for direct studies of gene regulation. The regulation of early genes appears to remain intact, and we have used extracts of these embryos to identify and to characterize a factor that interacts with the gut-specific *ges-1* control region (2, 8, 14).

Protocol 4.   Production of FUdR-blocked embryos

The basic protocol (see *Protocol 1*) for growing wild-type worms beginning with dauer larvae is modified as follows.

Equipment and reagents

- NGM (see *Protocol 1*) containing 40 µg/ml thymidine (Sigma) and 50 µg/ml fluorodeoxyuridine (Sigma)
- NGM–1 × egg tray (see *Protocol 1*, Part A, but use 100 ml of blended heated chicken egg/ tray)
- Microscope fitted with birefringence optics
- Alkaline hypochlorite solution (see *Protocol 3*)
- 30% sucrose solution
- 50 ml plastic, screw-cap, centrifuge tubes
- PBS (see *Protocol 1*)
- Liquid nitrogen and safety equipment

Method

1. Seed each FUdR–NGM (1 × egg) tray with about 1.5×10^6 dauer larvae. Incubate the trays at 20°C for 4–5 days. Check the status of growth every day to verify that the worms have exited dauer stage and are full of embryos.

2. View a sample of these worms using birefringence optics to verify that embryos intensely express gut granules[a] (see *Figure 1* of ref. 2).

3. Harvest and clean the adults as described above in *Protocol 1*, Parts C and D.

4. After the final rinse following sucrose flotation, dissolve the adult worms with alkaline-hypochlorite as described above in *Protocol 3*, steps 3–5.

5. Clean the embryos by 30% sucrose flotation (see above) in several 50 ml plastic screw-cap centrifuge tubes.

6. Transfer the embryos to one (tared) 50 ml tube and wash three times

---

Protocol 4. *Continued*

in PBS (see *Protocol 1*, Part D). After the final wash, weigh the wet pellet. Resuspend the embryos in an equal volume of PBS.

7. Immediately freeze the embryos by dripping into liquid nitrogen (see *Protocol 1*). Store at–70 °C.

[a] An adult worm grown on FUdR retains dozens of embryos, all blocked with a uniform and distinctive morphology: a region in the middle of these blocked embryos corresponds to the presumptive gut and gives the embryos a diagnostic 'fried egg' appearance under the dissecting microscope.

The advantages of the above procedure are that it is simple, robust, and tens of grams of tightly synchronized embryos can be achieved with little effort. Gene regulation seems intact in these blocked embryos: *ges-1* was expressed intensely but the *pho-1* gene (which is normally expressed after the point of arrest) was not. This is an obvious limitation for workers interested in late embryonic genes. Furthermore, there is a natural reluctance to use drug-blocked embryos, and if this reluctance can not be overcome we direct the readers to ref. 15 for the description of a method to produce transcriptionally active extracts from wild-type (unblocked) embryos isolated from adults grown in liquid culture. Similar methods have been used to characterize a protein complex involved in *C. elegans* dosage compensation (16). Whether factors are identified in FUdR-blocked embryos or in wild-type unblocked embryos, this identification is just the first in a series of steps required to establish that the identified factor is indeed the same factor that functions *in vivo*.

7. Production of unfertilized oocytes

We developed a procedure to produce unfertilized oocytes, based on the convenient properties of the temperature-sensitive, sperm-defective mutation *fer-1(b232)* (7). Both nuclear and cytoplasmic extracts were prepared from these oocytes and used to test for the presence of sequence-specific DNA binding factors (2). This method has been considerably extended by Aroian *et al.* (3) to purify actin-binding proteins present in oocytes. We briefly describe our method of producing the oocytes; improvements introduced in ref. 3 are noted.

Protocol 5. Production of unfertilized *C. elegans* oocytes from *fer-1(b232)* worms

Equipment and reagents
- One NGM–0.5 × egg tray (see *Protocol 1*, Part A)
- Two NGM–1 × egg tray (see *Protocol 4*) *without* thymidine and FUdR
- 30% sucrose
- 50 ml, screw-cap, plastic centrifuge tubes
- Sonicator with 4 mm diameter probe
- 44 μm and 21 μm Nytex filters
- Liquid nitrogen and safety equipment

Method

1. Grow *fer-1* (*b232*) worms on an NGM–0.5 × egg tray, as described in *Protocol 1*, until they form dauers; keep the incubation temperature at 16 °C. Harvest and purify the dauers as described in *Protocol 2*.

2. Add about 1.5 × 10^6 *fer-1* dauers to a seeded NGM–1 × egg tray (without thymidine and FUdR) and incubate at 26 °C for 3–4 days. Make sure the temperature is accurate and constant to within several tenths of a degree (if trays are incubated at 25 °C, larvae will be present and are almost impossible to separate from the oocytes).

3. Before harvesting, verify that the worms are full of oocytes by squeezing a small sample under a coverslip—oocytes look round and brown; no oval, refractile, fertilized embryos should be visible (see *Figure 6* of ref. 2).

4. Wash adult worms containing the oocytes off the trays with water.

5. Wash a further three times with water by centrifuging at 1600 *g* for 5 min, followed by flotation on 30% sucrose (see *Protocol 1*). Repeat the water washes three more times.

6. Aliquot 5–10 ml (maximum) of the worm pellet into a 50 ml tube and make up the volume to 25 ml with water. Keep the tube in ice and release oocytes from the worms with a 30-sec pulse of sonication at 35–40 watt output using the 4-mm diameter probe. Be cautious: it is very easy to apply too much power; on the other hand, if the power is not high enough, few oocytes will be released.[a]

7. Pour the whole suspension on to a 44-μm Nytex filter suspended over a 500 ml beaker to remove large debris. Wash the oocytes through the filter with water from a squirt bottle. Wash additional oocytes through by gently swishing the filter in a Petri dish full of water. Allow the filtered oocytes to settle under gravity in a 250 ml conical tube kept on ice; remove the supernatant by aspiration. Repeat this gravity settling step several times to remove any (more slowly settling) larvae that happen to be present. Repeat the above procedures with the rest of the pelleted worms.

8. Collect the oocytes on a 21-μm Nytex filter suspended over a Petri dish to remove small debris; wash the collected oocytes with water from a squirt bottle. Finally, wash the collected oocytes off the filter and centrifuge at about 100 *g* for 1 min at 4 °C; resuspend the pellet in an equal volume of PBS and freeze immediately by dripping into liquid nitrogen.[b]

---

[a] In the procedure of Aroian *et al.* (3), the sonication release of oocytes has been replaced by chemically induced expulsion using serotonin and levamisole, a procedure that appears to be both more rational and more controllable than sonication.

[b] The Aroian *et al.* procedure (3) describes variations on the above two filtering steps introduced with the aim of producing both greater quantities of oocytes and cleaner preparations.

8. Methods of worm disruption

We have prepared total extracts from worms in amounts ranging from a few mg (to use in genetic screens that depend on isoelectric focusing (6, 17)) up to several hundred grams (to purify digestive enzymes by conventional biochemistry (5, 6)). The method and extraction conditions will obviously depend on the use to which the extract will be subjected, and we offer here only rough guidelines based on our own experience. Preparation of nuclei (embryonic and oocyte) and preparation of nuclear extracts will be described in Section 9.

For the largest scale production of total worm extracts (usually from mixed-stage populations of wild-type worms), we used a motorized Stansted Cell Disruptor (Energy Services Co., Washington, DC), a rare and somewhat exotic instrument that can process hundreds of grams of worms relatively smoothly and at a suitably low temperature. A more widely available apparatus for large-scale worm disruption is a French Press; the use of this has been described in refs 18–20.

Sonication is a routine and flexible method for disrupting worms. We use a Braun-Sonic 2000; tip diameters of the interchangeable probes range from 4 to 19 mm. With the 4-mm tip, worms can be disrupted in a volume of less than 100 μl; with the large tip (and perseverance) up to 50 grams of worms have been disrupted (6). The sample is kept immersed in an ice-water bath, with sonication pulses (usually 30 sec) interspersed with periods of cooling (usually twice the length of the sonication pulse). The progress of the disruption is conveniently followed by viewing a small sample by dark-field microscopy. As a very rough guide, we have found that worms are disrupted by a total energy input of 35 watt-min/ml of sample.

A few further hints for successful worm disruption can be offered. It is our impression that worms that have been frozen are disrupted more easily than are fresh worms. To ensure efficient disruption, we generally aim for a ratio of 5–10 volumes of extraction buffer to 1 volume of worms. Frozen worm pellets can be melted directly in ice-cold extraction buffer, but this takes several minutes. If more rapid disruption is required (for example, for large-scale RNA extraction), the frozen pellets should first be pulverized in liquid nitrogen, either using a pre-cooled mortar and pestle or (our preferred method) by wrapping the pellets in a plastic envelope and smashing to a fine powder with a hammer (17). For the larger scale extractions (especially when using the motorized cell disruptor or similar apparatus), 10% glycerol is included in the solution to slow down settling of the worms and to make the suspension more uniform.

The addition of protease and nuclease inhibitors is an all-important consideration (see ref. 18). In the large-scale extraction of GES-1 (a serine esterase), the addition of standard inhibitors of serine proteases had to be avoided, but, somewhat surprisingly, proteolysis was never a problem (5). This is not always the case, and in the nuclear extraction procedure, described below, a collection of protease inhibitors is added. For large-scale RNA extraction (best done by the rapid dissolution of worms into a chaotropic salt solution (17), the addition of 1 mM ATA (aurintricarboxylic acid) provides an effective and very cheap nuclease inhibitor. However, this must be removed (for example, by an oligo-dT column) prior to adding any nucleic acid-modifying enzyme, such as reverse transcriptase.

Subsequent processing steps of the crude total extracts will obviously vary with the goals of the experiment. Total worm extracts are turbid and can be clarified by low-speed centrifugation (5, 6), by adjusting the pH to 5, and/or by adding $CaCl_2$ to 10–50 mM. The appropriateness of these steps in a particular procedure may only be determined by experience. Total extracts of worms grown as described above are very lipid-rich; after centrifugation, the floating skin can conveniently be removed by pouring the supernatant through several layers of (buffer-wetted) Miracloth (Calbiochem).

9. Preparation of nuclei and nuclear extracts from embryos and oocytes

We have used the following procedure primarily with FUdR-blocked embryos, with the aim of identifying DNA binding factors involved in the production of the embryonic gut (2, 8). However, we have also used the same basic procedure with oocytes and with unblocked embryos.

Protocol 6. Preparation of nuclei and nuclear extracts from embryos and oocytes

Equipment and reagents

- Wheaton stainless-steel tissue grinder; clearance 0.0005 inches (12.5 μm)
- 2 × Nuclear preparation buffer (2 × NPB): 20 mM Hepes pH 7.6, 20 mM KCl, 3 mM $MgCl_2$, 2 mM EGTA, 0.5 M sucrose, 1 mM DTT, 1 mM PMSF, 20 μM E-64 (Sigma)
- 1 × NPB
- Detergent buffer: 1 × NPB, 0.25% NP-40, 0.1% Triton X-100
- Glycerol
- Dark-field and fluorescence microscopes

- Nuclear extraction buffer (NEB): 20 mM Hepes pH 7.6, 350 mM NaCl, 2 mM EDTA, 25% glycerol, 0.5 mM DTT, 0.5 mM PMSF, 10 μM E-64
- DAPI (4′,6-diamidino-2-phenylindole)
- 15 ml centrifuge tubes
- Liquid nitrogen and safety equipment
- Dounce homogenizer and pestle A
- Equipment and reagents for a standard dye-binding assay

Method

1. Thaw 4–5 ml of embryos at room temperature; once thawed, place immediately on ice. Perform all subsequent steps on ice.

2. Transfer embryos to a pre-chilled Wheaton stainless-steel tissue grinder.[a] Add an equal volume of ice-cold 2 × NPB and homogenize

---

Protocol 6. *Continued*

with approximately 5–10 strokes (maximum of 15); (this requires considerable strength). Be careful not to over-homogenize as nuclei can begin to break. View a small sample under the dark-field microscope to check that the embryos have indeed been disrupted. Check the state of the nuclei by adding DAPI (4′,6-diamidino-2-phenylindole) to 1 µg/ml to a small aliquot of the nuclear suspension and examining using the fluorescence microscope.

3. Pour the homogenate into a 15 ml centrifuge tube, rinse the tissue grinder with a small amount of cold 1 × NPB, combine with the first homogenate, and pellet the nuclei at 4000 *g* for 5 min at 4 °C. (Save the supernatant as a crude cytoplasmic extract; to clarify, spin at 35 000 *g* for 30 min at 4 °C, carefully remove the supernatant avoiding the white scum at the top, add glycerol to 25% (v/v), quick freeze in liquid nitrogen, and store at −70 °C.)

4. Clean up the nuclear pellet by resuspending in 5 ml of ice-cold detergent buffer.[b] Transfer to a pre-chilled Dounce homogenizer and resuspend with five strokes of pestle A. Pellet unwanted debris at 100 *g* for 5 min at 4 °C. Transfer the supernatant containing the nuclei to a new tube. Recover trapped nuclei by washing the pellet two or three times with ice-cold detergent buffer.

5. After the final wash, spin the combined supernatants at 4000 *g* for 5 min at 4 °C to pellet the nuclei. Resuspend the pellet in 5 ml 1 × NPB and check an aliquot in the presence of 1 µg/ml DAPI on the fluorescence microscope. Re-pellet the nuclei as above.

6. Proceed with the nuclear extraction by resuspending the pelleted nuclei in 2–4 volumes (usually approximately 400 µl) of NEB. Transfer to a microcentrifuge tube and extract by gentle rotation for 45 min at 4 °C.

7. Spin at 10 000 *g* (in a microcentrifuge) for 15 min at 4 °C to pellet unwanted debris. Carefully remove the supernatant representing the crude nuclear extract. Use the crude extract immediately (preferably), or distribute it into small aliquots (say 20 µl), quick-freeze in liquid nitrogen, and store at −70 °C. Measure the protein concentration of the crude extract using a standard dye-binding assay.

[a] A normal glass Dounce homogenizer will not break open the embryos.
[b] 0.35% of either one of the separate detergents may work equally well.

Ref. 21 describes the isolation of nuclei to investigate nuclease digestion patterns of both active and inactive heat-shock genes. Nuclei active in 'transcription run-on' assays have been prepared both from *C. elegans* (11, 22) and from *Ascaris suum* (23). As noted above, the reader is referred to ref. 15 for the preparation of *in-vitro* transcription extracts from *C. elegans* embryos, and to ref. 16 for the use of embryonic nuclear extracts to study complexes of chromosomal proteins involved in dosage compensation.

10. The use of nuclear extracts to investigate DNA–protein interactions

As noted earlier, there have been few investigations of the interaction between specific DNA sequences and *C. elegans* proteins in nuclear extracts. One promising direction for the future is to add specific proteins produced in bacteria to transcriptionally active embryo extracts, using templates whose behaviour has been defined by transformation and genetics (15). Our own efforts have centred on proteins in embryonic nuclear extracts that bind to the promoter region of the gut-specific *ges-1* gene (2, 8) and, more recently, on proteins that interact with the promoter of the gene encoding the gut-specific GATA factor *elt-2* (14). Despite considerable efforts, DNA footprinting using embryonic nuclear extracts has rarely yielded clear results. Presumably the incomplete site-protection obtained reflects the fact that the extracts necessarily contain proteins from all cells of the embryo, not just the gut, and hence it is difficult to produce conditions where the template is saturated with binding protein. We have tried variations on the standard footprinting approach, for example using hydroxyl-radical protection or separating bound from unbound probe on native gels following the reaction (24); although protected regions could be discerned, the degree of protection was never impressive. The gel electrophoretic mobility shift assay (or band shift), rather than footprinting methods, probably represents the simplest, most sensitive, and most robust technique with which to explore DNA–protein interactions in nuclear extracts of *C. elegans*. The following represents the protocol we currently use in the laboratory, but many variations are available (25).

To produce double-stranded oligonucleotides suitable for band-shift assays, we have usually end-labelled one oligodeoxynucleotide using T4 polynucleotide kinase and [γ-^{32}P]ATP and then annealed with a 10-fold excess of its unlabelled complement. In this way, we minimize the interaction between trace amounts of labelled single-stranded oligonucleotide and a single-stranded DNA binding factor that is prominent in crude extracts (2).

The native gels used to separate bound from free probe are a standard formulation of 4–6% acrylamide (total:bis = 29:1) and half-strength Tris–borate–EDTA running buffer. The gels we use (BioRad Mini-PROTEAN) are 7 cm long by 0.15 cm thick, run in the cold room (with buffer pre-cooled in ice) at 100 V. We usually pre-run the gels for 15–30 min at 100 V.

A 5 × concentrated stock of binding buffer contains: 125 mM Hepes pH 7.6, 250 mM KCl, 5 mM DTT, 50 μM ZnSO$_4$, 1 mg/ml BSA, 50% glycerol, and 0.5% NP-40. These binding conditions work well with GATA factors, but may not be optimal for other classes of transcription factor.

The following represents a typical binding reaction:

Double-stranded oligonucleotide probe ($\sim$ 5 ng; 10^4–10^5 c.p.m.)	1 µl
5 $\times$ Buffer (see above)	4 µl
1% Orange G loading dye	1 µl
'Extract'	at least 1 µl
Poly-dIdC (0.1 or 1 µg/µl stock)	1 µl
Water to a total of 20 µl	

We have found that, under certain conditions, a dye such as Bromophenol blue can apparently disrupt a binding interaction; the dye Orange G does not appear to do this. Orange G migrates more rapidly than Bromophenol blue; if the gel is stopped just before the Orange G emerges, unlabelled probe will remain in the gel.

The above volumes are meant only as a rough guide. Poly-dIdC can be increased up to several micrograms per reaction to reduce 'non-specific' binding. The stock poly-dIdC is heated at 90°C for 10 min just prior to its addition, to reverse network formation. The reactions are assembled on ice, transferred to a room temperature water bath for 20 min and returned to ice; 18 µl of each reaction is loaded on the gel as soon as possible. After electrophoresis, gels are placed on Whatman # 1 filter paper, dried under vacuum (80°C for 1 h) and exposed to X-ray film or phosphorimager screens for times ranging from several minutes to overnight.

11. Manipulation of the living *C. elegans* embryo

Although the primary focus of this chapter has been on conventional biochemistry, the chapter will be closed by describing two techniques for manipulating the living *C. elegans* embryo. In recent years, these have been particularly powerful experimental tools for investigating cell fates and how these cell fates are influenced by neighbouring cells.

11.1 Blastomere culture

A procedure to investigate the cellular basis of early *C. elegans* development has been described in detail by Edgar (26). Early embryos are isolated from hermaphrodites, treated with hypochlorite and chitinase to remove the resistant eggshell, and then stripped of their underlying vitelline membrane with a very fine glass pipette. When cultured in embryonic growth medium (26, 27), these devitellinized embryos undergo the majority of embryonic development and are now permeable to drugs such as cytochalasin or aphidicolin. By separating early blastomeres (with an eyelash or micromanipulator), the cellular requirements of various developmental events can be determined. For example, Goldstein (28) used this technique to show that contact between the P_2

cell and the EMS cell in the four-cell embryo is both necessary and sufficient to induce gut differentiation in EMS. Thorpe *et al.* (29) extended this technique by combining blastomeres of different genotypes to determine which genes are required in the signalling-cell P_2, and which are required in the receiving-cell EMS, during gut induction.

11.2 Laser ablation

A more general method of examining the roles of individual cells during all stages of *C. elegans* development is laser ablation, described in detail by Bargmann and Avery (30). This technique takes advantage of the transparency of *C. elegans*; a laser is focused on the nucleus of the cell of interest and repeatedly pulsed until 'damage' (scarring or nuclear breakdown) is observed. The ablated cell may divide once or twice more, but further development is inhibited. Laser ablation does not eliminate the operated cell, and thus cannot prevent immediate interactions. However, it has been extremely useful for studying interactions between individual blastomeres (e.g. see ref. 31) and, by killing all early blastomeres but one, for identifying the products of the remaining blastomere (32).

With the foundation provided by the techniques of blastomere culture and laser ablation (as well as genetics), current investigations are now able to approach the molecular basis of cell fate and intra-embryonic signalling.

References

1. Nelson, G. A., Roberts, T. M., and Ward, S. (1982). *J. Cell. Biol.*, **92**, 121.
2. Stroeher, V. L., Kennedy, B. P., Millen, K. J., Schroeder, D. F., Hawkins, M. G., Goszczynski, B., and McGhee, J. D. (1994). *Dev. Biol.*, **163**, 367.
3. Aroian, R. V., Field, C., Pruliere, G., Kenyon, C. and Alberts, B. M. (1997). *EMBO J.*, **16**, 1541.
4. Lewis, J. A. and Fleming, J. T. (1995). In *Methods in cell biology*, Vol. 48 (ed. H. F. Epstein and D. C. Shakes), p. 1. Academic Press, San Diego, CA.
5. McGhee, J. D. (1987). *Biochemistry*, **26**, 4101.
6. Beh, C. T., Ferrari, D. C., Chung, M. A., and McGhee, J. D. (1991). *Dev. Biol.*, **147**, 133.
7. Argon, Y. and Ward, S. (1980). *Genetics*, **96**, 413.
8. Hawkins, M. G. and McGhee, J. D. (1995). *J. Biol. Chem.*, **270**, 14666.
9. Riddle, D. L. (1988). In *The nematode* Caenorhabditis elegans (ed. W. B. Wood), p. 393. Cold Spring Harbor Laboratory Press, NY.
10. Wood, W. B. (ed.) (1988). *The nematode* Caenorhabditis elegans. Cold Spring Harbor Laboratory Press, NY.
11. Schauer, I. E. and Wood, W. B. (1990). *Development*, **110**, 1303.
12. Egan, C. R., Chung, M. A., Allen, F. L., Heschl, M. F. P., Van Buskirk, C. L., and McGhee, J. D. (1995). *Dev. Biol.*, **170**, 397.
13. Fukushige, T., Schroeder, D. F., Allen, F. L., Goszczynski, B., and McGhee, J. D. (1996). *Dev. Biol.*, **178**, 276.

14. Fukushige, T., Hawkins, M. G., and McGhee, J. D. (1998). *Dev. Biol.*, **198**, 286.
15. Lichtsteiner, S. and Tjian, R. (1995). *EMBO J.*, **14**, 3937.
16. Chuang, P. T., Lieb, J. D., and Meyer, B. J. (1996). *Science*, **274**, 1736.
17. McGhee, J. D., Birchall, J. C., Chung, M. A., Cottrell, D. A., Edgar, L. G., Svendsen, P. C., and Ferrari, D. C. (1990). *Genetics*, **125**, 505.
18. Epstein, H. F. and Liu, F. (1995). In *Methods in cell biology*, Vol. 48, (ed. H. F. Epstein and D. C. Shakes), p. 437. Academic Press, San Diego, CA.
19. Aamodt, E. J. and Culotti, J. G. (1986). *J. Cell Biol.*, **103**, 23.
20. Lye, R. J., Porter, M. E., Scholey, J. M., and McIntosh, J. R. (1987). *Cell*, **51**, 309.
21. Dixon, D. K., Jones, D., and Candido, E. P. M. (1990). *DNA Cell Biol.*, **9**, 177.
22. Honda, S. and Epstein, H. F. (1990). *Proc. Natl Acad. Sci. USA*, **87**, 876.
23. Cleavinger, P. J., McDowell, J. W., and Bennett, K. L. (1989). *Dev. Biol.*, **133**, 600.
24. Schroeder, D. F. (1998). Ph D Thesis, University of Calgary, Canada.
25. Ausubel, F. M., Brent, R., Kingston, R. E., Moore, D. D., Seidman, J. G., Smith, J. A., and Struhl, K. (1987–1998). *Current protocols in molecular biology*. Wiley,
26. Edgar, L. G. (1995). In *Methods in cell biology*, Vol. 48 (ed. H. F. Epstein and D. C. Shakes), p. 303. Academic Press, San Diego, CA.
27. Shelton, C. A. and Bowerman, B. (1996). *Development*, **122**, 2043.
28. Goldstein, B. (1992). *Nature*, **357**, 255.
29. Thorpe, C. J., Schlesinger, A., Carter, J. C., and Bowerman, B. (1997). *Cell*, **90**, 695.
30. Bargmann, C. I. and Avery, L. (1995). In *Methods in cell biology*, Vol. 48 (ed. H. F. Epstein and D. C. Shakes), p. 225. Academic Press, San Diego, CA.
31. Hutter, H. and Schnabel, R. (1994). *Development*, **120**, 2051.
32. Mello, C. C., Draper, B. W., Krause, M., Weintraub, H., and Priess, J. R. (1992). *Cell*, **70**, 163.

12

Conventional genetics

JONATHAN HODGKIN

1. Introduction

Caenorhabditis elegans was originally chosen as an experimental organism partly because of the advantageous features of its genetic system (1). Genetic analysis is a particularly powerful approach for investigating complex biological topics. One gene, once thoroughly characterized, can become the starting point for further exploration. With a strain bearing a mutation in the gene, functional interactions between genes/gene products can be examined by generating double mutants through simple crosses with previously characterized mutants. Genetic screens can be performed for mutations with similar phenotypes or for mutations that suppress or enhance a phenotype. The power of the genetic system in *C. elegans* means that these experiments can move very quickly, and novel genes thereby identified can be mapped and cloned.

Normally, populations of this animal consist almost entirely of diploid self-fertilizing hermaphrodites, which each produce a limited number of sperm and a larger number of oocytes. Both types of gametes result from conventional meioses, with about one crossover event per chromosome. The advantages of self-fertilization are several: first, mutants with very severe behavioural or anatomical phenotypes can nevertheless grow and reproduce as homozygous stocks, because the animal needs little more than a functional gut and pharynx in order to ingest and digest bacterial food. Second, self-fertilization means that a hermaphrodite carrying a recessive mutation will automatically produce animals homozygous for that mutation, as one-quarter of its progeny, according to standard Mendelian principles. As a result, screening for mutants is made much easier, because there is no need to set up specific crosses in order to generate homozygotes. Diploidy also means that lethal or sterile mutations can be maintained, by using heterozygous animals. Finally, self-fertilization means that both recessive and dominant modifier mutations can be easily detected, if screens for suppressors or enhancers of a given phenotype are carried out. Again, there is no need to set up specific crosses; the screening can be carried out on whole populations.

Making mutants is therefore easy, but it would be hard to proceed further in genetic analysis without the ability to exchange genetic material between individuals. In conventional *C. elegans* genetics, this is made possible by the existence of a male sex, which can cross-fertilize hermaphrodites, contributing sperm which then preferentially fertilize the hermaphrodite's eggs.

This chapter describes standard procedures for selfing and crossing; mutagenesis and mutant screening; characterization of mutants; gene mapping; temperature-shift experiments; and mosaic analysis.

2. Reproduction, maintenance, and scoring

C. elegans hermaphrodites are able to reproduce either by self-fertilization, using sperm that have been made during larval development, or by cross-fertilization, using sperm contributed by a male after mating. Hermaphrodites are unable to cross-fertilize each other. Once a hermaphrodite has mated with a male, the sperm from the male are used preferentially over the sperm made by the hermaphrodite.

2.1 Selfing

Many procedures for *C. elegans* genetics involve only self-fertilization. In a standard self-cross, a single L4 hermaphrodite of known genotype is picked on to a small seeded NGM plate (see Chapter 4). An L4 stage animal is used because this ensures that the animal has not been mated, and permits collection of the entire self-progeny brood. If the brood is to be scored, the hermaphrodite should be picked to a fresh plate at daily intervals until no more eggs are produced (about 3 days at 20°C). Daily transfer ensures that the entire brood can be scored without any possibility of confusion between F_1 and F_2 progeny.

2.2 Crossing

Crosses between hermaphrodites and males are carried out by placing a few hermaphrodites, either L4 or virgin young adults, together with an excess of young adult males (e.g. three hermaphrodites and five males, numbers may be varied depending on fecundity) on a small seeded plate. The bacterial lawn should not touch the edge of the plate because males tend to swim up the plastic wall and die of desiccation. Males are particularly liable to do this if no adult hermaphrodites are present on the plates, so the hermaphrodites should be transferred to the crossing plate before the males. Old or starved males mate poorly or not at all.

Cross-progeny and self-progeny are often distinguished by using hermaphrodites homozygous for a recessive visible mutation, so that all self-progeny are marked and all cross-progeny are unmarked. Alternatively, complete outcrossing can usually be guaranteed by picking a single hermaphrodite at L4 stage and placing it on a small plate with five or six adult males. Another method is to 'purge' hermaphrodites of their own sperm, by transferring them to fresh plates until they cease to produce eggs and lay a few unfertilized oocytes. These purged hermaphrodites usually remain competent to be cross-fertilized for several days more. However, the purging procedure takes several days, and sometimes the resulting aged hermaphrodites have become irreversibly sterile.

Single-pair matings (i.e. one hermaphrodite and one male) are sometimes necessary and can result in cross-progeny in over 90% of crosses under good conditions. Efficiency of single-pair matings can be increased by using males from the Californian wild isolate CB4855, which exhibit markedly higher male fertility than N2 males (2).

Frequently, it is necessary to transfer a marker using a male. A general problem in *C. elegans* genetics is that males expressing a mutant phenotype are often impaired in mating or unable to mate at all. Males may express a mutant phenotype because they are homozygous or hemizygous for a recessive mutation, or heterozygous for a dominant mutation. Impaired males can, however, often mate successfully if their partner hermaphrodites also have a mutant phenotype, caused by a *dpy* or *unc* mutation, that causes them to move at least as slowly as the impaired males. Alternatively, for most recessive mutations, heterozygous XO males (*m/+*) can be used to transfer a particular marker, *m*. For sex-linked mutations that prevent mating, for example *unc-1*, heterozygous males can sometimes be made by using an X chromosome duplication that carries a wild-type allele of the marker in questions: *unc-1/0; mnDp66*. Alternatively, heterozygous XX males can be constructed using *tra* mutations. The best mutation for this purpose is the unusual *tra-2* allele *q276:* XX animals which are homozygous for this autosomal mutation, on LGII, are phenotypically male and can cross-fertilize hermaphrodites, although less efficiently than wild-type males. In all crosses with heterozygous males, only half the cross-progeny will receive the marker of interest, so it is necessary to pick several (usually four to six will be enough) cross-progeny hermaphrodites to separate plates, and type them by progeny testing, in order to identify those of the desired genotype.

Hermaphrodites of some mutant phenotypes are self-fertile, but are difficult to cross-fertilize because of the abnormality or absence of a vulva. In these cases, crossing large numbers of animals together is often effective (for example six crossing plates, each with 10 hermaphrodites and 20 males). In extreme cases, successful mating can be made possible by creating an artificial vulval opening with a micropipette, or by employing artificial insemination.

2.3 Source of males for crossing

Males (XO) are found only rarely (0.2%) in populations of self-fertilizing hermaphrodites (XX), as a result of spontaneous X chromosome loss. If the males are capable of successful mating, then a male stock can be established and maintained by picking these rare males and mating them with sibling hermaphrodites. The male frequency in the next generation will be 50%, if mating occurs efficiently.

Male stocks are best maintained by picking a few hermaphrodites and a larger number of males to a fresh plate at each generation. If this is not done, then the male frequency falls in successive generations, because reproduction by selfing is more efficient than reproduction by crossing.

It may be convenient to maintain a wild-type (N2) male population in sustained exponential growth at 15°C, by daily transfers, so that healthy young adult males are always available for crossing.

These wild-type males can then be crossed with hermaphrodites homozygous for any autosomal recessive mutation, i.e. to generate cross-progeny males heterozygous for this mutation. These males can then be used for further crosses, in order to transfer the mutation into a new genotype.

Homozygous males can also be generated directly from a homozygous hermaphrodite stock, or heterozygous males can be generated directly from a balanced heterozygous hermaphrodite stock. There are two possible methods: one is to increase the spontaneous male frequency (normally 0.2%) by heat-shock treatment. In this procedure, 10–20 hermaphrodites at mid-L4 stage are incubated for 6 h at 30°C on a small seeded NGM plate, and then selfed at 20°C. Their progeny will usually include 2–5% males, although the efficiency of this procedure is variable, depending on the mutant stock used. The second method is to construct a stock homozygous for a *him* (High Incidence of Males) mutation (3). These mutations result in the increased loss of X chromosomes during hermaphrodite gametogenesis, so self-progeny male frequencies are increased to 20% or more. The most useful mutations for this purpose are *him-5(e1490) V* and *him-8(e1489) IV*, both of which result in male frequencies above 30%.

2.4 Maintenance of lethal or sterile mutations with genetic balancers

Most *C. elegans* mutants, even those with severe anatomical or behavioural defects, are able to reproduce as homozygous hermaphrodites, by means of self-fertilization, and can therefore be maintained as stable stocks. However, about one-third of the genes in this organism are essential, so mutations that eliminate their activity lead to a lethal or sterile phenotype. Such mutations must therefore be maintained as heterozygotes, using a wild-type allele of the gene in question (*let/+*). A population founded by a single hermaphrodite of this genotype will rapidly be taken over by wild-type homozygotes (*+/+*), and the lethal will be lost, unless some precautions are taken.

The simplest kind of balancer is a tightly linked visible marker in *trans* to the lethal (*let +/+ unc*), for example a strong *unc* mutation, which will permit stable propagation except for rare recombination events. If flanking markers are used (*dpy* + *unc/*+ *let* +), such recombination events will be even rarer. Ideally, however, lethals should be maintained using a balancer that suppresses recombination. There are three kinds of balancer routinely used in *C. elegans* genetics:

(a) Duplications of the region of interest, either free duplications or duplications attached to another chromosome. For example, *eDp6* is a free chromosomal duplication, which carries most of the genes on the right arm of LGIII, and behaves as an unstable extra chromosome. Any lethal in this region can be maintained as a strain of genotype *let/let; eDp6*. Another example is *mnDp10*, which duplicates the right end of the X chromosome and is attached to LGI, so it is transmitted in Mendelian proportion. Some duplications can recombine with their parent chromosome, and are therefore unsuitable balancers.

(b) Intrachromosomal rearrangements (most probably inversions), which prevent recombination within the rearranged region. For example, *mnC1* can be used as a stable balancer for any mutation between *dpy-10* and *unc-52* on LGII, which amounts to about two-thirds of this chromosome.

(c) Translocations between two chromosomes will act as balancers for most or all of the translocated regions. Two popular translocations are *eT1(III; V)* and *nT1(IV;V)*. The first balances the right arm of LGIII and the left arm of LGV; the second balances most of LGIV and LGV.

Good balancers are available for most, but not all, the *C. elegans* genome. A skeleton genetic map is shown in *Figure 1*, with the well-balanced regions indicated by a dashed line. A list of commonly used balancers is given in *Table 1*, and a more extensive description can be found in ref. 4.

2.5 Scoring crosses

Broods can be scored most effectively by physically removing the worms from the plate, usually by means of a platinum-wire wormpick. If many worms of the same phenotype need to be counted, it may be more convenient to count and remove them by aspiration, using a Pasteur pipette connected to an aspirator or vacuum pump. The end 2 cm of the pipette should be bent almost at right angles to the barrel, and the tip flamed so that it is rounded and slightly narrowed. Such a pipette can be used to suck up worms efficiently without breaking the agar surface of the plate.

2.6 Analysing crosses

All crosses involve the generation of progeny by the fusion of eggs and sperm, both of which result from meiotic divisions. In a complex cross, there will be multiple possible gamete types contributed from each parental germline, and

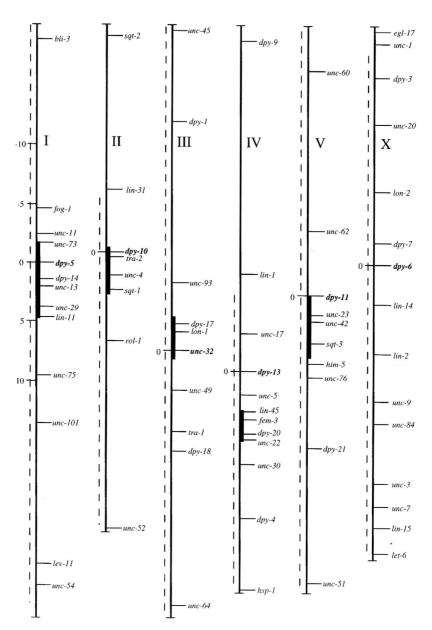

Figure 1. Skeleton genetic map, showing some of the genes most commonly used for mapping. Each of the six chromosomes is drawn with the left (or minus) end up and the right (or plus) end down. Genetic map coordinates are measured in centiMorgans (1 cM = 1% recombination), relative to an arbitrarily chosen zero point marker in the approximate centre of each chromosome. This marker is shown in bold type for each chromosome. Genes to the left of this marker are given negative coordinates and genes to the right are given positive coordinates. On each autosome, the central cluster region, with higher gene density, is indicated by a thicker line. The dashed lines indicate regions for which good genetic balancers are available (see *Table 1*).

Table 1. Commonly used balancers

Name	LG	Region balanced[a]	Description
hT1(I;V)	I, V	LE to +2.6, LE to 0	Reciprocal translocation
hT2(I;III)	I, III	LE to +15, −2 to RE	Reciprocal translocation
sDp2(I;f)	I	LE to +1.5	Free duplication
hIn1	I	+9.5 to +27.2	Inversion, homozygous viable
mC6	II	−5 to +8	Carries dpy-10; homozygous. viable
mnC1	II	0 to +23	Carries dpy-10, unc-52; homozygous viable
mnDp34(II;f)	II	+3 to +23	Free duplication
sC1	III	−27 to −9	Carries dpy-1, homozygous viable
sDp3(III;f)	III	−19 to −0.3	Free duplication
qC1	III	−19 to +7	Carries dpy-19, glp-1, mog-1; homozygous sterile
eT1(III;V)	III,V	−0.3 to RE, LE to +1	Reciprocal translocation; breaks in unc-36; homozygous viable
eDp6(III;f)	III	+5.4 to RE	Free duplication
nT1(IV;V)	IV,V	−5 to RE, LE to +9	Reciprocal translocation; homozygous vulvaless
szT1(I;X)	I, X	LE to +2.1, −14 to RE	Carries lon-2; dominant Him
mnDp10(X;I)	X	+7 to RE	Attached duplication
mnDp3(X;f)	X	+11 to RE	Free duplication
mnDp1(X;V)	X	+17 to RE	Attached duplication

[a]Abbreviations: LE, left end; RE, right end

consequently a variety of progeny types. It is often helpful to tabulate all possible resultant genotypes and phenotypes, using the Punnett Square method. Examples are given in *Figures 2* and *3*. This ensures that all progeny are accounted for, and simplifies the prediction of their expected frequencies.

3. Mutagenesis and screening

3.1 Mutagenesis

The vast majority of mutations generated in *C. elegans* have been produced by treatment with EMS (ethyl methanesulfonate), which is a convenient and potent mutagen for this organism. The standard protocol for EMS muta-genesis is shown in *Protocol 1*. This uses a high dose of EMS, about 50 mM, so the possibility of multiple mutations is increased, but for most purposes the advantage of an increased mutation frequency outweighs the disadvantages of extraneous mutational background. A lower dose (down to 5 mM) can be used if necessary. EMS generates predominantly GC to AT transitions, so not all possible mutational changes can be easily generated with this mutagen. However, it also produces other kinds of point mutation and deletions, albeit at a lower frequency.

SPERM

		Non-recombinant 1 - p		Recombinant p	
		n +	+ *d*	*n d*	+ +
Non-recombinant gametes,	*n* +	N	WT	N	WT
frequency = 1 - p	+ *d*	WT	D	D	WT
Recombinant gametes,	*n d*	N	D	ND	WT
frequency = p	+ +	WT	WT	WT	WT

EGGS

Frequency of phenotype N   = (1-p)(1-p)/4 + 2p(1-p)/4

Frequency of *n d* / *n* + recombinants   = 2p(1-p)/4

Figure 2. Punnett Square illustrating the segregation of phenotypes in the self-progeny from a *trans*-heterozygous hermaphrodite, genotype *nbm-1* +/+ *dpy-5*. The recessive mutant alleles *nbm-1* and *dpy-5* are abbreviated to *n* and *d*, respectively, and the corresponding wild-type alleles are written +. The four possible phenotypes are WT (Wild-type), N (Nbm), D (Dpy), and ND (Nbm Dpy). In the absence of linkage (independent segregation, $p = 0.5$), these would be produced in the ratio 9:3:3:1. For tight linkage ($p \sim 0$), the ratio would be approximately 2:1:1:0.

Protocol 1.   EMS mutagenesis

Caution: EMS is carcinogenic and mutagenic, and must only be handled in a fume hood, with gloves. It is effectively hydrolysed and inactivated by strong alkali, so all contaminated glassware must be allowed to soak in at least 1 M NaOH for 24 h or more.

Equipment and reagents

- Wild-type (N2) hermaphrodite worms grown on a 9 cm plate by the usual plate culture method (see Chapter 4, *Protocol 2*). The population of worms should be healthy with plenty of mid-to-late larval stages.
- M9 buffer (see Chapter 4, *Protocol 7*)

- EMS (ethyl methanesulfonate) (Sigma, cat. no. M-0880)
- 2 M NaOH
- NGM plates seeded with OP50 (see Chapter 4, Section 3.3)
- Wide-bore (2 cm) glass test-tube[a]

Method

1. Wash the worms off the culture plate using 5 ml M9 buffer.

2. Centrifuge briefly (30 sec, 500 g) and resuspend the worms in 10 ml of M9 buffer.

3. Re-centrifuge and resuspend the worms in 1 ml of M9 buffer.

4. In a fume hood, place 3 ml of M9 buffer into a wide-bore (2 cm) glass test-tube[a] and add 20 µl EMS. Swirl the tube until the EMS has dissolved.

5. Add the suspension of worms in buffer (1 ml) to the EMS solution (this concentration of EMS corresponds to 47 mM). Swirl to mix and then leave to incubate at room temperature (22 °C) for 4 h. Swirl the tube approximately every 30 min to prevent the worms becoming too anoxic.

6. Allow the worms to settle at the bottom of the mutagenesis tube, and remove as much as possible of the supernatant. Add the supernatant to an excess volume of 2 M NaOH to destroy the EMS.

7. Wash the worms three times with at least 10 ml of M9 buffer each time, discarding the supernatant by adding it to excess 2 M NaOH.

8. Take up the worms in a minimal volume (0.2 ml or less) of M9 buffer and add this to a 9 cm plate containing an NGM/OP50 lawn.

9. When this has soaked in, pick actively moving L4 animals to a fresh plate, and allow them to recover and mature into young adults (about 24 h).

10. From this population, pick single healthy adults, each to a separate 9 cm plate.

11. Allow egg-laying to proceed for a day, then transfer each adult to a new plate.

12. Screen the F_2, and possibly also the F_1, generation for mutant phenotypes.[b]

13. If a plate becomes starved, wash off the population with a small volume of M9 buffer, and add to fresh plates for re-feeding.

---

[a] Glassware is preferable to plasticware because worms tend to stick to various plastics, and can become damaged and therefore more susceptible to directly toxic effects of the mutagen.
[b] Maximal numbers of mutations seem to derive from the second and third days of egg-laying by the treated hermaphrodites. 50–100 eggs per day are laid after mutagenesis, each of which will develop into an F_1 hermaphrodite. Rare dominant mutations will be expressed in the F_1 generation, but most mutations are recessive, so they will not be detected until the F_2 generation. The complete F_2 brood from 50 F_1 worms will amount to more than 10000 F_2 individuals, which can be screened for mutants. With so many worms on each plate, the bacterial food is soon exhausted, and it is much harder to detect mutants in a starving population, which necessitates the re-feeding step 13.

SPERM

		Non-recombinant 1 - p		Recombinant p	
		n *d*	+ +	*n* +	+ *d*
Non-recombinant gametes, frequency = 1 - p	*n* *d*	ND	WT	N	D
	+ +	WT	WT	WT	WT
Recombinant gametes, frequency = p	*n* +	N	WT	N	WT
	+ *d*	D	WT	WT	D

(EGGS, shown vertically on the left)

Frequency of phenotype N or D $= p^2/4 + 2p(1-p)/4$

Frequency of phenotype ND $= (1-p)^2/4$

Frequency of phenotype WT $= (3 - 2p + p^2)/4$

Figure 3. Punnett Square illustrating the segregation of phenotypes in the self-progeny from a *cis*-heterozygous hermaphrodite, genotype *nbm-1 dpy-5/+ +*. Conventions as in *Figure 2*. For tight linkage ($p \sim 0$), the four possible phenotypes WT, N, D, ND would be produced in the approximate ratio of 3:0:0:1.

A variety of other chemical mutagens have been employed, for which protocols can be found elsewhere (5). The most useful mutagens for achieving gene knockouts are those creating small deletions (0.1–5 kb), and for this purpose it is likely that the optimal agent is trimethylpsoralen/UV.

Protocol 2. Psoralen mutagenesis[a]

Equipment and reagents

- 1 M 4,5',8 trimethylpsoralen (TMP, trioxsalen; Sigma, cat. no. T6137) in DMSO. This is equivalent to 0.228 g TMP per ml. Store in the dark.
- Wild-type (N2) hermaphrodite worms (see *Protocol 1*)
- M9 buffer (see Chapter 4, *Protocol 7*)
- Unseeded and seeded NGM agar plates (see Chapter 4, *Protocol 2* and Section 3.3 therein)
- Long-wave UV source (360 nm is effective)

Method

1. Wash wild-type (N2) hermaphrodite worms using M9 buffer. (See *Protocol 1*, steps 1–3.)

2. Add 67 μl 1 M TMP to 2 ml of the worm suspension. Allow to stand in the dark for 15–60 min.

3. Take up the worms, which will have collected at the bottom of the tube, in as small a volume as possible and apply the worms to an unseeded NGM plate. Allow excess TMP to soak into the plate.

4. Irradiate the plate using the long-wave UV source. The optimal dose is best determined empirically, by a kill-curve: good mutation rates are obtained at 10% lethality (egg death). For a calibrated source, use approximately 340 μW/cm^2 for 60 sec.

5. Wash off the worms and transfer them to a fresh seeded plate.

6. Keep this plate in the dark for at least 5 h longer because the treated worms remain UV-sensitive for some time. Then transfer individuals or populations to NGM plates as desired.

[a] Modified from ref. 6.

For generating chromosomal rearrangements (duplications, deficiencies, translocations, and inversions), ionizing radiation is more effective than EMS. Both gamma-irradiation and X-irradiation have been widely employed. For gamma-irradiation, a dose of 750–7500 R is delivered from a ^{60}Co source, at a dose rate of 50 R/sec. For X-irradiation, the standard high dose is 5000–7500 R, using an X-ray source delivering 50–650 R/min. Washed worms are irradiated directly, on the surface of an unseeded NGM plate.

For generating transposon-induced mutations, two effective mutator loci, *mut-2* and *mut-7*, have been employed (7; and Plasterk, personal communication). Use of strains carrying either of these mutators, which lead to germline transposition of Tc1 and other transposons, is preferable to using Bergerac strains such as RW7000. The Bergerac strains carry a high transposon load, and germline transposition appears to occur at a lower rate than with the mutator loci. Mutagenesis by transposition is occurring all the time in any of these strains, so populations can be directly screened for mutants.

3.2 Mutant screens and selections; clonal screens

For many purposes, mutants can be recovered simply by the mass screening of an F_2 population derived from a small number of mutagenized P_0 hermaphrodites. This population can also be subjected to a variety of manipulations, for example drug treatments, in order to enrich or select for a particular mutant phenotype.

In some cases, a mutant phenotype can only be detected by means of a treatment which is lethal to the worms, for example an enzyme assay. In this situation, some form of clonal screen is appropriate. This is usually done by picking individual hermaphrodites from the F_2 generation after mutagenesis to separate plates, each of which will produce a clonal line of descendants. These separate populations can then be sampled and subjected to whatever test is desired. The same approach can be used when screening for phenotypes that can only be detected in populations or by testing multiple individuals, for example when searching for mutants defective in male mating. After EMS mutagenesis, between 1 in 1000 and 1 in 10000 F_2 clones is homozygous mutant for a given gene (depending on the target size of that gene). Therefore, clonal screens are labour-intensive but feasible, and have often been used with great success. Note that a significant fraction (20% or more) of superficially wild-type F_2 hermaphrodites will carry mutations that prevent the establishment or propagation of a clonal line, due to sterility or other deleterious phenotypes.

Handling such a large number of populations poses some logistical problems. Microtitre well plates have been used to grow up multiple clones. However, it is important to use those constructed with separate wells, otherwise worms are liable to crawl from one well to the next and cross-contaminate the clones. Alternatively, clones can be grown on smaller plates (3.5 cm), but these do not offer much advantage over the normal small plates (5 cm or 6 cm), and may be less convenient to pour, handle, and store.

3.3 Isolation of lethal mutations

Lethality and sterility are phenotypes that are easily detected, but they obviously present problems in propagating the mutations responsible. One solution for analysing essential genes is to search for conditional mutations in such genes, usually temperature-sensitive (t.s.) mutations. F_2 mutant clones are established as above and grown at the permissive temperature, and then samples from each population are shifted to the restrictive temperature, to see if they are still viable and fertile. T.s. lethals obtained in this way are very useful, and can be used for temperature-shift experiments and suppressor screens (see below). However, t.s. mutations are rare and sometimes very hard to find. Also, a t.s. mutation may have residual function even at restrictive temperature, and therefore give a misleading idea of the null phenotype for the gene concerned.

A more general solution is to search for lethal mutations using a genetic balancer. For example, the balancer chromosome *mnC1* balances most of LGII, including the convenient visible marker *unc-4*. A set of lethal mutations in this region was isolated by mutagenizing *unc-4* hermaphrodites and crossing them with *unc-4/mnC1* males. F_1 non-Unc progeny hermaphrodites were picked to separate plates, and their self-progeny (F_2) screened for the absence of adult Unc-4 animals (about 5% of broods). These hermaphrodites must have carried a *let* mutation linked to *unc-4*, and balanced by *mnC1: unc-4 let-x/mnC1*. Each lethal was therefore stably balanced from the start (8).

Variations on this type of protocol have been used to isolate large numbers of lethals in several different parts of the genome. To some extent, the ease of such lethal screens depends on the properties of the balancer used.

4. Cleaning and characterizing mutants

4.1 Mutant isolation

In general, for homozygous viable mutations, single hermaphrodite worms expressing the mutant phenotype of interest should be picked from F_2 populations to separate plates. No more than one mutant of a given type should be retained from each population founded by a single mutagenized P_0, in order to ensure independence. Usually a mutant will give rise to a self-progeny brood, all of which express the parental mutant phenotype. Occasionally wild-type, or heterogeneous, progeny are produced. This can happen for a variety of reasons: the picked animal may have been a phenocopy, or it may have been genetically mosaic in its germline, or the phenotype may be dominant or variable. In any case, further mutants (if any) should be picked from the progeny and selfed, until a stable mutant line can be established.

All mutant lines should preferably be established by selfing from a single homozygous hermaphrodite, or a single heterozygous hermaphrodite, in the case of recessive lethal or sterile mutants. Establishment from a single individual reduces the genetic heterogeneity in the resulting population, but does not eliminate it. Mutagenized stocks inevitably carry a large burden of mutations in addition to the one of interest. It can be estimated that the standard treatment with EMS induces one lethal mutation on 50% of all chromosomes, as well as a much larger number of less deleterious or silent changes. Mutagenesis treatments with other agents, such as ionizing radiation or transposons, are also likely to generate additional mutations. Therefore, it is desirable to remove as many as possible of these background mutations, which may exert significant modifier effects, by repeated outcrossing against wild type. For homozygous viable mutations, a single mutant should be picked and crossed with wild-type males to yield phenotypically F_1 wild-type heterozygous hermaphrodites, which are then selfed to give homozygous mutant F_2 progeny. Ideally, this process should be repeated several times, but in practice many mutant stocks have only been outcrossed once, or not at all, so one should always be wary of possible background effects. Nevertheless, even after a single outcross, the re-isolated mutant will carry wild-type chromosomes for most of the rest of the genome, which will be either heterozygous or homozygous, so further growth will tend to dilute away deleterious unlinked mutations—another advantage of the selfing mode. Beneficial modifiers, on the other hand, will be selected and become homozygous during continued growth. Spontaneous mutations may also arise and modify the phenotype, but fortunately the spontaneous mutation rate in N2 is very low. To avoid the accumulation of modifiers, it is advisable to freeze (see Chapter 4) and store a reference strain for each mutant, as soon as it has been outcrossed.

Outcrossing will fail to remove tightly linked mutations. Also, mutations generated in *trans* to balancer chromosomes, and maintained over balancers, will have a larger load of permanently linked background mutations, which will remain unless explicit crosses are carried out to remove them.

Moreover, screens for lethal mutations are often carried out using a visible marker, usually one of the standard *unc* or *dpy* mutations used for mapping purposes. Lethals are generated on the same chromosome as the visible marker, which may significantly affect the phenotype of the lethal, especially since these markers often confer very abnormal phenotypes, either morphologically or behaviourally. Consequently, it is always preferable to examine the phenotype of a new mutation in the absence of linked markers.

Finally, if it is important to create isogenic lines, known to differ from wild-type only at the locus of interest, then ideally one should construct a triple mutant, flanking the mutant locus with closely linked markers on either side, and then remove these flanking markers in further crosses.

4.2 Mutant characterization

A simple outcross between a mutant hermaphrodite and a wild-type male will yield information about the basic properties of the mutation responsible for the mutant phenotype: recessive or dominant, autosomal or sex-linked.

Additional characterization, in terms of anatomical and developmental description, lineage analysis, cellular and subcellular analysis, biochemical and molecular properties, can be pursued indefinitely, and methods for such analyses are covered in chapters elsewhere in this volume. However, there are a number of basic tests that can be applied to any mutant, which will affect its further investigation, and may reveal important, useful, or illuminating properties. Some of these are described below, assuming that one is dealing with a homozygous viable, recessive mutant. These tests can be appropriately modified to deal with dominant or lethal mutations. The following is a useful checklist:

- *Penetrance and expressivity*: Is the mutant 100% penetrant? That is, are all homozygotes distinguishable from the wild type? Is the expressivity constant: that is, do all homozygotes exhibit essentially the same phenotype? To some extent, all mutants are more variable in phenotype than wild-type, but some exhibit extreme variability, so that genotypically identical individuals can vary from apparent normality to complete inviability. Obviously, it will be harder to work with such highly variable mutants.

- *Maternal effects*: Is the phenotype of a homozygote derived from a homozygous mutant hermaphrodite parent the same as that for a homozygote derived from a heterozygous parent? Sometimes the phenotype will be significantly weaker in the latter case, because a wild-type gene product has been contributed to the offspring. This is particularly true for many mutations affecting embryogenesis.

- *Gene dosage effects*: If a mutation has been mapped, and a genetic deficiency that spans the locus is available, the phenotype of a homozygous mutant (*m/m*) can be compared to that of a hemizygous mutant (*m/Df*). For mutations that cause only partial loss of function, the hemizygote may exhibit a stronger or different phenotype. Also, some genes are haploinsufficient, so *m/+* heterozygous animals may be discernibly different from wild-type ones.

- *Expression during the life-cycle*: Is the phenotype visible throughout the life-cycle, and does it vary in strength? Behavioural phenotypes, in particular, can be very different between the L1 stage and the adult, because the L1 nervous system is significantly different from that of later stages. Also, behaviours such as egg-laying can only be scored in the adult.

- *Expression in the two sexes*: Stocks are usually maintained only as hermaphrodites, so examining the phenotype of a male usually requires additional crosses, or the use of a *him* mutant, as above. However, the male phenotype can be very informative, both in terms of behaviour, which is substantially more complex than in the hermaphrodite, and in terms of anatomy. The male tail is modified into an elaborate copulatory organ, and the correct development of this depends on many determinative and morphogenetic events. Consequently, male tail anatomy provides a sensitive assay for normal post-embryonic development. The ability to mate successfully provides a sensitive assay for normal behaviour.

- *Hermaphrodite fertility*: It is often useful to count the entire brood produced by selfing a single mutant hermaphrodite. Wild-type worms produce about 330 ± 30 mature progeny at 20°C, with 0.2% males and fewer than 1% unhatched eggs. Significant increases in embryonic lethality may indicate a low-penetrance developmental defect, and increases in male frequency may reveal a meiotic defect.

- *Male fertility*: Subtle defects in male development or mating behaviour can be revealed by inefficient mating. In the standard male fertility assay (2), a single male is picked at L4 stage, and added to a plate with six young adult females. A day later, he is transferred to a fresh plate with six new females, and so on for successive days until no more progeny are sired. Total progeny are counted. To generate females (that is, animals with hermaphrodite anatomy and oogenesis but no spermatogenesis), the standard mutation is *fem-1(hc17)*. This is a temperature-sensitive mutation, so XX *hc17* hermaphrodites, grown at 15°C, are shifted to 25°C: all their progeny will mature into fertile females.

- *Temperature effects*: Is the mutant phenotype the same at low (15 °C) and high (25 °C) temperature, as compared to that seen at the standard growth temperature (20–22 °C)? Many mutations exhibit weak temperature-sensitive effects, and some are strongly t.s. In general, 'temperature-sensitive' means 'heat-sensitive', with a stronger phenotype at elevated temperature. Cold-sensitive mutations are rarer, but not unknown. In the case of a strong effect, it becomes feasible to carry out temperature-shift analysis, to test for time of gene product action. Also, if a mutant is inviable or nearly inviable at restrictive temperature, it becomes easy to carry out powerful selections for revertants or extragenic suppressors.

- *Dauer formation*: Is the mutant capable of forming dauer larvae on starvation? Also, does maturation via the dauer stage affect the phenotype? Dauers can easily be purified from a starved culture by washing with 1% SDS, which rapidly kills all stages except eggs and dauer larvae. Larval survivors are rinsed with M9 buffer, and added to fresh bacterial lawns, whereupon they will exit the dauer stage and resume growth as L4 larvae.

- *Starvation effects*: Is the phenotype affected, either enhanced or suppressed, by starvation?

- *Response to suppressors*: Is the mutant phenotype affected by informational suppressors? There are two kinds of suppressor in common use. One class is amber suppressors, which permit readthrough of amber (UAG) stop codons (9). Most amber mutations result in complete or almost complete loss of function, so amber-suppressibility usually indicates a severe mutation. The other class is *smg* suppressors, which prevent degradation of messages that contain premature stop codons (10). Some mutants are suppressed by *smg* suppressors, but some are enhanced.

4.3 Temperature-shift analysis

A temperature-sensitive mutation of a gene can be used to determine when, during development, a particular gene product is required. Usually 25 °C is used as the restrictive temperature, and 15 °C as the permissive temperature. For genes active during post-embryonic development, it is convenient to synchronize animals at hatching. They are then transferred from 25 °C to 15 °C (down-shift) or vice versa (up-shift), at defined times thereafter. Alternatively, for analysis of earlier development, individuals can be synchronized by dissecting gravid hermaphrodites, and picking embryos at the 2-cell stage, which can be readily recognized with a good dissecting microscope. Development takes twice as long at 15 °C as at 25 °C, and times are usually normalized to 25 °C rates when designing and reporting temperature-shift experiments.

4.4 Mosaic analysis

Mosaic analysis provides information about where a given gene must be active in order to permit normal development. Expression studies, carried out by means of reporter genes, *in-situ* hybridization, or immunofluorescence studies, can provide information about where a given gene is expressed, and where its products are located, but they do not reveal which are the essential sites of expression. Examination of genetically mosaic animals, in which some cells are wild type and some cells are mutant for a given gene, can provide this information. Mosaic analysis will also reveal if a particular gene is cell-autonomous, or cell-non-autonomous, in its action. An additional use of genetic mosaics is to examine the role of maternal-effect genes, by constructing germ-line chimeras. These are hermaphrodites that have lost a particular gene-activity in the germline, but retained it in the soma. Progeny produced by such an animal will lack both maternal and zygotic gene activity, and may exhibit a different phenotype from those which only lack the zygotic gene activity—for example early embryonic as opposed to larval lethality.

Genetic mosaics of *C. elegans* are usually generated by making use of un-stable chromosomal duplications or extrachromosomal transgene arrays (11). Such elements can be lost during mitotic divisions, at a low frequency ($< 0.5\%$ per division), as compared to normal chromosomes, which are essentially never lost during mitosis. A zygote that is chromosomally mutant for a given gene, but carries a wild-type copy of this gene on a duplication, genotype *m/m; Dp(+)*, can therefore give rise to a mature animal in which a small fraction of the cells have lost the duplication, and are therefore mutant, *m/m*. If the gene is needed in these cells, then a mutant phenotype will be seen.

To interpret observations on such a mosaic, it is essential to know where and when the duplication was lost. The invariant cell lineage of *C. elegans* provides a major advantage in this context, because the final pattern of cells with and without the duplication can usually be used to pinpoint the exact division at which the loss occurred.

The distribution of cells with and without the duplication must be deter-mined by using some kind of additional marker associated with the duplica-tion. This marker must be cell-autonomous in action, and ideally it should be possible to score in most or all cells. Until recently, the most useful general marker has been the gene *ncl-1*. Most cells that are mutant for *ncl-1* have enlarged nucleoli, which can be clearly recognized in living animals by means of Nomarski microscopy. A zygote of genotype *ncl-1/ncl-1; Dp(ncl-1(+))* can give rise to an adult animal consisting mostly of normal cells with small nucleoli, and a few cells which have lost the duplication and therefore have large nucleoli. This gene is also convenient because it is located on *sDp3*, a well-characterized duplication with suitable loss frequency. In order to create a compound duplication carrying the gene of interest as well as *ncl-1(+)*, it is possible to fuse an extragenic array or duplication carrying this gene with *sDp3*, by means of gamma-irradiation. After irradiation with 3000–4000 R, 0.2–1% of progeny should carry a fused duplication, which can be recognized by means of appropriate chromosomal markers. Alternatively, a mixed extragenic array can be constructed by co-injecting a clone for the gene of interest, and a *ncl-1*(+) clone. Such arrays often exhibit the same kind of somatic instability as free duplications, and can therefore be used in the same way.

An alternative cell marker is provided by a *sur-5::GFP* fusion, which confers bright nuclear fluorescence on almost all somatic cells (but not germline cells) (12). This has the advantage that the marker can be detected much more easily than can the *ncl-1* marker, and populations of animals can be rapidly screened for mosaics by using a good dissecting microscope equipped with epifluorescence optics. Loss of an array carrying the fusion during development will give rise to an individual with patches of dark nuclei. Once a mosaic animal has been identified, it can be examined in more detail.

The detection and interpretation of mosaic animals can be facilitated in a variety of other ways, depending on the properties of the gene in question. For a thorough description of the methods available, and the possible problems that may be encountered, see ref. 12.

Genetically mosaic hermaphrodites can sometimes be used to generate progeny that lack both maternal and zygotic contributions for a particular gene. For example, animals homozygous for a mutation in the gene *gpb-1,* encoding a G protein beta subunit, are inviable, dying at the first larval stage. These mutant animals can be rescued by an extragenic array carrying *gpb-1*(+). A small fraction of the mutant hermaphrodites carrying the array are mosaic, such that they retain the array in the soma but have lost it in the germline. These animals are often fertile, but lay eggs which exhibit abnormal early cleavages and fail to develop, thereby revealing an early embryonic function for this gene product (13).

5. Genetic mapping procedures

For all forward genetics in *C. elegans,* starting from novel mutant phenotypes and proceeding to genetic and molecular characterization, it is essential to map the mutations responsible. In the absence of map information, there is no guarantee that the trait under investigation is genetically determined, rather than resulting from some environmental or stochastic variable. A map position provides simple proof of a genetic basis. Moreover, much of the genetic characterization outlined in Section 4 can only be carried out easily if a gene has been mapped, and can be manipulated using linked markers. Finally, a genetic map location is an essential first step in the positional cloning of any gene.

Most of the *C. elegans* genetic map has been assembled using visible markers, but the use of polymorphic DNA sequences as mapping markers has become an attractive alternative. The next four sections discuss conventional approaches for mapping, using visible markers, and the following two sections describe sequence-based methods.

5.1 Assignment of linkage

Mapping starts with the assignment of a gene to a linkage group. Strategies for mapping a mutant will vary, depending on the phenotype involved. For example, assume that a new behavioural mutant has been isolated, and that initial outcrossing has shown that it is an autosomal recessive, *nbm-1*. For mapping a behavioural trait, it is convenient to use morphological markers, such as *dpy*, *sma*, or *lon*. There are *dpy* markers in the cluster regions of each of the five autosomes (see *Figure 1*), so each can be tested. About 80% of autosomal mutations map within or close to these clusters, and therefore the probability of detecting linkage is high. The mutant (*nbm-1/nbm-1*) is crossed with heterozygous males carrying a *dpy* marker, say *dpy-5* on chromosome I (*dpy-5/+*), and F_1 hermaphrodite cross-progeny are picked singly to separate plates, and allowed to self. Half of these will be *dpy-5/+; nbm-1/+*, recognizable because they will segregate WT (wild-type), Dpy, and Nbm F_2 progeny, as shown in *Figure 2*. If the double-mutant phenotype, Nbm Dpy, can be recognized, and is present as approximately 1/16 of F_2 progeny, then it is unlikely that *nbm-1* is linked to *dpy-5*. However, it is often difficult to recognize or score the double mutant, so a less ambiguous procedure is to pick single Nbm F_2 hermaphrodites to separate plates, and ask what fraction of them produce 1/4 Nbm Dpy F_3 progeny. In the absence of linkage, 2/3 will do so, but if the two genes are linked, then a much lower fraction will do so. The expectation for linkage p is $2p/(1+p)$, so if the two genes are 5 cM apart, then the fraction will be 0.09. That is, only 1 in 11 Nbm hermaphrodites will segregate Nbm Dpy progeny, as a result of recombination between the two loci (see *Figure 2*).

5.2 Standard 2-factor mapping

Measurement of the recombinational distance of a new gene from one or two known markers will usually permit accurate placement on the genetic map. This will also confirm linkage, and avoid the pitfall of incorrect linkage assignment, which can arise from ambiguous 3-factor data. Most 2-factor distances used in the construction of the genetic map have been derived by segregation from *cis*-heterozygotes, because in this method reliable numbers can be derived by counting only a few complete broods, and no progeny testing is entailed. The method can be illustrated by pursuing the example initiated above, with a hypothetical behavioural mutation, *nbm-1*, mapped to LGI. The linkage test will have yielded a doubly homozygous line, *nbm-1 dpy-5*. Hermaphrodites from this line are crossed with wild-type males, to yield wild-type F_1 hermaphrodites of genotype *nbm-1 dpy-5/+ +*. These hermaphrodites are selfed, and the complete F_2 self-progeny broods of one or more are scored. As shown in *Figure 3*, the ratios of the four possible progeny phenotypes (WT, Dpy, Nbm, Dpy Nbm) are determined by the linkage, p, between *nbm-1* and *dpy-5*. Calculation is usually based on the fraction of recombinant phenotypes (Dpy + Nbm) to the total number of all four phenotypes; this is given by $(2p-p^2)/2$, which is approximately equal to p, when p is small.

In some cases, it is not possible to distinguish two of the progeny types, or they cannot be counted accurately, or they are lethal or subviable, and therefore under-represented in the brood. In all these cases, it is sufficient to count the other two phenotypes. In this example, if it was impossible to distinguish Dpy and Dpy Nbm, then only WT and Nbm would be counted. The fraction Nbm/(Nbm + WT) is given by $(2p-p^2)/3$.

A variety of other types of crossing and scoring can be employed if necessary—for example, when dealing with dominant mutations, or with phenotypes that can only be recognized in the male sex. About 20 different calculations have been used for the various special circumstances, and can be found in ACeDB (14). However, most of them are rarely used, so they are not described here.

5.3 3-factor and multipoint crosses for ordering genes

In standard 3-factor crosses, a gene can be ordered relative to two known markers, placing it unambiguously on the map. To illustrate this, consider the *nbm-1* example further. Suppose this has been assigned to LGI, and linkage to the *dpy-5* marker has been measured, giving a distance of approximately 2 cM ($p = 0.02$). It could therefore lie either to the left or to the right of *dpy-5*. In order to determine this, another marker on LGI is used, *unc-13*, which is located about 2 cM to the right of *dpy-5*. A double-mutant hermaphrodite, *dpy-5 unc-13*, is crossed with males carrying *nbm-1*, and F$_1$ progeny hermaphrodites of genotype *dpy-5 + unc-13/+ nbm-1 +* are identified by progeny testing. The F$_2$ progeny will consist mostly of WT, Nbm, Dpy Unc, as well as rare Dpy non-Unc and Unc non-Dpy, resulting from recombination events between *dpy-5* and *unc-13*. These rare recombinants are picked and selfed, to determine whether they carry the *nbm-1* mutation. If 10 recombinants of each type were picked, and it was found that 10/10 Dpy recombinants carried *nbm-1*, and 1/10 Unc recombinants carried *nbm-1*, then one could conclude that the order of the three genes is *dpy-5 nbm-1 unc-13*, with *nbm-1* much closer to *unc-13*. If, however, 0/10 Dpy recombinants and 10/10 Unc recombinants carried *nbm-1*, then one could assume that *nbm-1* is probably to the left of *dpy-5*. This is not a rigorous conclusion, because it might still be located just to the right of *dpy-5*. The linkage measurement makes this unlikely, but often the confidence limits on linkage estimates will be large, so that the linkage value may be misleading. Therefore, to be sure, another 3-factor cross should be carried out, using a marker to the left of *dpy-5*.

In this example, it might be difficult to recognize the Nbm phenotype in a Dpy or Unc background, so testing the recombinants by selfing would not be possible. In such circumstances, the recombinants would instead be typed by crossing with a male carrying *nbm-1*. The Nbm phenotype could then be recognized in the non-Dpy or non-Unc cross-progeny.

This example also illustrates the utility of 3-factor crosses in constructing tightly linked double mutants. If *nbm-1* was located only 0.1 cM from *unc-13*, then the double mutant *nbm-1 unc-13* would be very hard to construct, starting with a 2-factor cross with the genes in *trans (nbm-1+/+unc-13)*, but it would be easy to construct by means of this 3-factor cross.

5.4 Mapping by means of deficiencies and duplications

A large part of the *C. elegans* genome is now covered by genetic deficiencies or duplications. These can be used for gene mapping, and in favourable circumstances this approach is much easier and less ambiguous than using recombinational mapping. For example, if a new lethal mutation has been mapped to the vicinity of a standard *unc* marker, and there are available deficiencies which include the *unc* gene, then it is simple to test whether these deficiencies also include the *let* gene. Males of genotype *unc let/+ +* are crossed with hermaphrodites carrying each deficiency. If Unc cross-progeny are produced (*unc let/Df +*), then the deficiency does not include the *let* gene; if no Unc cross-progeny are seen, then the deficiency does include the *let* gene. Duplications can be tested in a similar manner, by setting up crosses to generate animals of genotype *unc let/unc let; Dp*.

The effectiveness of this approach depends on the availability of suitable deficiencies and duplications. Some parts of the genetic map have many, some have few. Also, one potential hazard is that some rearrangements have 'holes' in them, so that they delete (or duplicate) fewer genes than their apparent extents on the genetic map.

5.5 STS mapping by means of PCR

An ever-increasing number of polymorphic markers are available for the *C. elegans* genome, which can be used for direct mapping of any gene (15, 16). Most of these markers are precisely located in the sequenced genome, and are therefore STSs (sequence-tagged sites). Methods using STSs depend on PCR, and require molecular manipulations which are not needed in conventional mapping, but STS mapping has major advantages in terms of speed and flexibility. Ultimately, it is likely that the genome will be saturated with poly-morphic STSs, to a density of one every 100 kb or less, which will mean that any mutation can be rapidly mapped to this resolution simply by carrying out a series of PCR reactions.

Most, though not all, of these polymorphisms are insertions of the trans-poson Tc1. All such Tc1 insertions can be uniquely detected by PCR using general primers for Tc1, and specific primers in the adjacent sequence. The Bergerac strains (such as RW7000) of *C. elegans* contain several hundred Tc1 insertions that are absent from the Bristol strain, and which have provided the majority of these polymorphisms, but several other strains with activated Tc1 transposition have been used as sources of additional inserts. Also, Tc1 inserts are currently sparse or absent in some regions of the genome, and in these areas non-Tc1 restriction fragment length polymorphisms (RFLPs) have sometimes been identified, which can also be used for mapping.

Major advantages of STS mapping are that the markers are dominant and can be scored in any mutant background, even in a dead embryo, and that any number of different STS markers can be scored independently so long as they generate distinguishable PCR products.

To illustrate the method, suppose that a new embryonic lethal mutation has been identified, *let-x*, which needs to be mapped. It will initially need to be maintained by means of heterozygous hermaphrodites, *let-x/+*, which will produce at each generation one-quarter of dead eggs (*let-x* homozygotes). Once such a hermaphrodite is identified, by means of selfing, it can be crossed with a wild-type male, to yield male progeny. If half the cross-progeny are male, this demonstrates that the mutation is autosomal. If the viable cross-progeny are only one-quarter male, rather than one-half, then the mutation is sex-linked, and a different strategy must be employed (see below). For an autosomal mutation, take 10 or more male progeny (half of which will be carrying *let-x*), and mate them with 20 or more hermaphrodites of a Bergerac strain, usually RW7000. Animals of this strain are relatively slow-moving and slow-growing, so cross-progeny can be recognized by their improved movement. Pick 40 such hermaphrodites to separate plates, and identify those which segregate dead eggs of the Let-x phenotype. These hermaphrodites must therefore be *let-x/+*, and heterozygous for all Bergerac chromosomes. From these F_2 broods, pick approximately 70 individual dead eggs (arrested embryos) to separate microcentrifuge tubes, and carry out PCR reactions on them, as in *Protocol 3* (modified from ref. 15). For the first round of mapping, standard primers for each of the five autosomes are used (15). In the absence of linkage, three-quarters of the dead eggs will give a signal for a particular STS, because they are homozygous or heterozygous for the Bergerac site in question. In the presence of linkage, fewer than three-quarters will give a signal, depending on how tight the linkage is between *let-x* and the STS.

Once linkage has been established, a second round of mapping can be carried out using the same protocol, but this time using a set of primers specific just to the identified linkage group. Standard sets of about five STSs per linkage groups were defined (15) and have been widely used. The results from this set of tests should locate *let-x* to a resolution of 5 cM or better. The process can be repeated using further STSs within this region, if they are available, or with larger numbers, to improve the resolution.

If the lethal mutation turned out to be located on the X chromosome, a different starting procedure is used. Hermaphrodites of a Bergerac race (RW7000) are crossed with Bristol (N2) males, to yield male progeny that carry only a Bergerac X chromosome. These males are mated with *let-x/+*

hermaphrodites, identified by selfing as above. About 10 cross-progeny herm-aphrodites are picked to separate plates, and those which produce one-quarter Let-x dead embryos are identified. These must be *let-x/+*, and heterozygous for the Bergerac X chromosome. Individual dead eggs are then picked and tested with X-chromosome-specific STSs.

Procedures for constructing appropriate heterozygotes will obviously be simpler if the mutant of interest is homozygous viable, because the genetic manipulations will be easier and the PCR signals stronger. Also, many Tc1 transposons have now been identified and defined as STSs in non-Bergerac strains, which are much healthier. Males of these strains (e.g. DH424) will mate efficiently, unlike Bergerac males, so heterozygotes can be constructed by crossing the mutant of interest with homozygous DH424 males.

In principle, mutations can be very tightly located by using STS mapping alone, but in practice it is convenient to combine this with recombinational mapping using visible markers. For example, if a gene of interest is flanked by visible markers, then it is simple to pick recombinants that have separated the flanking markers, and test only those by PCR. As a result, only recombination events in or near the region of interest are tested. This will greatly reduce the number of reactions that need to be run.

Protocol 3. PCR analysis of single eggs or worms

Equipment and reagents

- Heterozygous (*let/+*) hermaphrodite worms
- Lysis buffer: 10 mM Tris–HCl pH 8.3, 50 mM KCl, 2.5 mM MgCl$_2$, 0.45% NP-40, 0.45% Tween-20, 0.01% gelatin. Autoclave and store at 4°C or freeze in aliquots at –20°C.
- 10 mg/ml Proteinase K in ddH$_2$O. Store in aliquots at –20°C.
- Lysis mix: 100 μl lysis buffer and 1 μl Proteinase K
- 10 × PCR buffer (Amplitaq)
- 15 mM MgCl$_2$

- PCR primers at 2 mM
- 100 × dNTP: 25 mM each of dATP, dCTP, dGTP, dTTP
- Amplitaq (Perkin Elmer–Cetus)
- 2-cm lengths of nylon fishing line (monofilament, 0.235 mm/10 lb test) with the ends cut at a 45° angle.
- Mineral oil
- Thermocycler
- Equipment and reagents for agarose gel electrophoresis

Method

1. Permit the heterozygous (*let/+*) hermaphrodites to lay eggs for 12 h or more.

2. Remove the hermaphrodites and incubate the plates for 24 h to permit viable eggs to hatch. The remaining unhatched eggs will be homo-zygous *let/let*. Alternatively, if the desired genotype is not embryonic lethal, simply select individual larval or adult worms of the appropriate phenotype.

3. Pick single dead (unhatched) eggs, or individual worms, using the nylon fishing line.[a] Transfer the filament and sample into a PCR tube

---

Protocol 3. *Continued*

containing 2.5 μl of the lysis mix. Check that the egg (or worm) has been successfully deposited in buffer, using a dissecting microscope.

4. Freeze the tube contents for 20 min or more, on dry ice.

5. After the sample has frozen, add a drop of mineral oil. Place the tube in the thermocycler and incubate at 60°C for 60 min to allow digestion. Then incubate at 95°C, 15 min, for heat inactivation.

6. *PCR analysis*: To the 2.5 μl lysate, add 2.5 μl 10 x PCR buffer, 2.5 μl 15 mM MgCl$_2$, 2.5 μl of each primer, 0.25 μl 100 x dNTP, 0.125 μl Amplitaq, and H$_2$O to 25 μl. Set up six to eight samples for each primer pair, using the following PCR profile:

(a) 95°C, 3 min

(b) 95°C, 40 sec

(c) 50°C, 90 sec

(d) 72°C, 2 min (or 1 min/kb)

Go to step (b) and repeat the cycle 35 times.

(e) 72°C, 10 min

(f) 6°C, hold.

7. Examine the products by agarose gel electrophoresis.[b]

[a] The use of fishing line was introduced by Eric Lambie.
[b] 5% acrylamide/TBE gels can be used instead of agarose.

5.6 PCR mapping of deficiencies

A gene may be located quite precisely by means of deficiency mapping, as described above, but the molecular extent of many deficiencies is known only approximately, or not at all. The ability to carry out PCR reactions on individual eggs or larvae means that it is possible to define the extent of deficiencies on the physical map, and thereby greatly narrow the physical region in which a gene of interest lies. Zygotes which are homozygous for a deficiency undergo extensive embryonic development before arresting, and may even hatch before dying at an early larval stage (in the case of some small deficiencies). These inviable zygotes contain several hundred nuclei, and can be used as a source of DNA for PCR. Suitable primer pairs are chosen from the genomic sequence in the region concerned, aiming for products between 100 and 500 bp, but check the primers on a wild-type genomic DNA preparation to confirm their suitability and compatibility. PCR reactions are carried out on sets of individual dead eggs or larvae, as in *Protocol 3*. If a signal is consistently observed, then the tested sequence does not lie within the deficiency. If a signal is not observed, then the sequence is probably deleted by the deficiency. However, it is important to be confident that the PCR has worked, so control primers from another part of the genome should be included, to avoid false-negatives. Also, false-positives can sometimes be obtained, particularly for large deficiencies and deficiencies that include haplo-insufficient genes. In these cases, some of the dead eggs or dead larvae will be heterozygous (*Df/+*) rather than homozygous (*Df/Df*), and will therefore give a misleading positive signal. Consequently, it is important to test sets of 5–10 individuals in parallel.

5.7 Molecular cloning of mapped genes

The application of these mapping procedures will identify the region of the physical map in which the gene lies. Apart from the effort applied, the resolution achieved depends on the quality of the physical map and the density of connections between the genetic and physical maps in the vicinity of the gene of interest. However, several YAC and many cosmid genomic DNA clones, from the physical map, can be screened for a gene by transformation rescue of the mutant phenotype or through molecular detection of the genetic alteration. This then returns the investigator to the genome sequence data that may have been the original stimulus to start working with *C. elegans*.

References

1. Brenner, S. (1974). *Genetics*, **77**, 71.
2. Hodgkin, J. and Doniach, T. (1997). *Genetics*, **146**, 149.
3. Hodgkin, J., Horvitz, H. R., and Brenner, S. (1979). *Genetics*, **91**, 67.
4. Edgley, M. L., Baillie, D. L., Riddle, D. R., and Rose, A. M. (1995). In *Methods in cell biology*, Vol 48 (ed. H. F. Epstein and D. C. Shakes), p. 147. Academic Press, San Diego, CA, USA.
5. Anderson, P. (1995). In *Methods in cell biology*, Vol 48 (ed. H. F. Epstein and D. C. Shakes), p. 31. Academic Press, San Diego, CA, USA.
6. Yandell, M. D., Edgar, L. G., and Wood, W. B. (1994). *Proc. Natl Acad. Sci. USA.*, **91**, 1381.
7. Collins, J., Saari, B., and Anderson, P. (1987). *Nature*, **328**, 726.
8. Sigurdson, D. C., Spanier, G. J., and Herman, R. K. (1984). *Genetics*, **108**, 331.
9. Kondo, K., Makovec, B., and Waterston, R. H. (1990). *J. Mol. Biol.*, **215**, 7.
10. Pulak, R. and Anderson, P. (1993). *Genes Dev.*, **7**, 1885.
11. Herman, R. K. (1995). In *Methods in cell biology*, Vol. 48 (ed. H. F. Epstein and D. C. Shakes), p. 123. Academic Press, San Diego, CA, USA.
12. Yochem, J., Gu, T., and Han, M. (1998). *Genetics*, **149**, 1323.
13. Zwaal, R. R., Ahringer, J., van Luenen, H. G. A. M., Rushforth, A., Anderson, P., and Plasterk, R. H. A. (1996). *Cell*, **66**, 619.

14. Eeckman, F. H. and Durbin, R. (1995). In *Methods in cell biology*, Vol. 48 (ed. H. F. Epstein and D. C. Shakes), p. 584. Academic Press, San Diego, CA, USA.
15. Williams, B. D., Schrank, B., Huynh, C., Shownkeeen, R., and Waterston, R. H. (1992). *Genetics*, **131**, 609.
16. Williams, B. D. (1995). In *Methods in cell biology*, Vol. 48 (ed. H. F. Epstein and D. C. Shakes), p. 81. Academic Press.

A1

List of suppliers

Becton Dickinson and Company, 1 Becton Drive, Franklin Lakes, NJ 07417. *Becton Dickinson Europe*, 38241 Meylan Cedex, France.

Bio101, Inc., 1070 Joshua Way, Vista, CA 92083, USA.

Braun *Branson Sonic Power Co.*, Eagle Road, Danbury, Connecticut, 06810. USA.

Dr. C. Bargmann, Box 0452, Department of Anatomy, University of California, San Francisco, CA 94143-0452, USA.

C. elegans, Web site (for valuable tips and discounts) (http://elegans.swmed.edu/).

Cadisch Precision Meshes Ltd., Arcadia Avenue, Finchley, London N3 2J7, UK.

Caenorhabditis Genetics Center, University of Minnesota, 250 Biological Sciences Center, 1445 Gortner Avenue, St Paul, MN 55108–1095, USA; fax: (612) 625–5754; e-mail: stier@biosci.cbs.umn.edu

Calbiochem, 10394 Pacific Centre Court, San Diego, CA 92121, USA.

Dr. A. D. Chisholm, Department of Biology, University of California, Santa Cruz, CA 95064, USA.

Clark Electromedical Instruments, (http://www.clark.mcmail.com/).

Energy Services Co., 1010 Vermont Avenue, N.W., Suite 500, Washington, D.C. 20005, USA.

Eppendorf, (http://www.eppendorf.com/).

Evergreen Scientifics, 2300 E. 49th Street, PO Box 58248, Los Angeles, CA 90058, USA. (800) 421-6261.

Falcon, Falcon products are available from Becton Dickinson and can be ordered from the addresses given above.

Fast Inc., (http://www.fastmultimedia.de/)

Dr. Andrew Fire, Carnegie Institute of Washington, Department of Embryology, 115 West University Parkway, Baltimore, MD 21210, USA.

Fisher Scientific UK, Bishop Meadow Road, Loughborough, Leicestershire LE11 5RG. (44) 01509 231166; fax (44) 01509 231893.

Fisher Scientific and Fisher Biotechnology, 9999 Veteran's Memorial Drive, Houston, TX 77038-2499, USA. (800) 766-7000.

Halocarbon Products Corp., P.O. Box 661, River Edge, NJ 07661, USA.

Hamamatsu, (http://hamamatsu.com/).

Dr. Min Han, Department of MCD Biology, University of Colorado, Boulder, CO 80309-0347, USA.

Dr. R. Horvitz, Department of Biology, Massachusetts Institute of Technology, Cambridge, MA 02139, USA.

ICN BioMedicals, Unit 18, Thame Park Business Centre, Wenman Road, Thame OX9 3XA, UK and 3300 Hyland Avenue, Costa Mesa, CA 92626, USA.

Improvision, UK, (http://improvision.com/).

Dr. Y. Jin, Department of Biology, University of California, Santa Cruz, CA 95064, USA.

Dr. J. Kramer, Department of CM Biology, Terry 8-703, Northwestern University Medical School, 303 East Chicago Avenue, Chicago, IL 60611, USA.

Kramer Scientific Corp., 5 Westchester Plaza, Elmsford, New York 10523, USA.

Leica, (http://www.leica.com/).

Medical System Corp., One Plaza Road, Greenvale, New York 11548, USA.

Midwest Scientific, P.O. Box 11750, St. Louis, Missouri 63105, USA. (800) 227-9997.

Dr. D. M. Miller, Department of Cell Biology, Vanderbilt University Medical School, 1161 21st Avenue South, Room U-2212, MCN, Nashville, TN 37232-2175, USA.

M. J. Research, Inc., P.O. Box 98549, Las Vegas, Nevada 89193, USA. (800) 729-2165.

Nalge Company, Box 20365, Rochester, N.Y. 14602, USA.

Narishige USA, Inc., 404 Glen Cove Avenue, Sea Cliff New York 11579, USA.

Narishige Europe Ltd., Unit 7, Willow Business Park, Willow Way, London SE26 4QP, UK.

Nunc

Nalge Nunc Int., 2000 North Aurora Road, Naperville, IL 60563-1796 USA.
Nunc A/S, P.O. Box 280, DK-4000 Roskilde, Denmark.

Olympus, (http://www.olympus.com/).

Perceptics, (http://www.perceptics.com/).

Dr. D. Pilgrim, G-507 Biological Sciences Building, University of Alberta, Edmonton, Alberta T6G 2E9, Canada.

Research Services Branch (RSB) of the National Institute of Mental Health (NIMH), part of the National Institutes of Health (NIH) (Web site: http://rsb.info.nih.gov/nih-image/about.html).

Dr. Ralph Schnabel, Institut Fuer Genetik, TU Braunschweig, Spielmannstr. 7, D-38023 Braunschweig, Germany.

Science Products GmbH, Hofheimer Str. 63, D-65719 Hofheim, Germany; fax: (49) 6192 901398; (49) 6192 901396.

Simi Gmbh, (http://www.simi.net)

Sony, (http://www.sony.de1/).
Sutter, (http://www.sutter.com/).
Wheaton Science Products, 1501 North 10th Street, Millville, NJ 08332, USA.
World Precision Instruments Corp., 175 Sarasota Center Blvd., Sarasota, FL 34240, USA.
Zeiss, (http://www.zeiss.com/).

Index

Index

Index